Elements of Paleontology
edited by
Brenda R. Hunda
Cincinnati Museum Center

CRITICAL HISTORY FOR TOMORROW'S PALEONTOLOGY

Victor Monnin
Archives Henri-Poincaré

Shaftesbury Road, Cambridge CB2 8EA, United Kingdom

One Liberty Plaza, 20th Floor, New York, NY 10006, USA

477 Williamstown Road, Port Melbourne, VIC 3207, Australia

314–321, 3rd Floor, Plot 3, Splendor Forum, Jasola District Centre, New Delhi – 110025, India

103 Penang Road, #05–06/07, Visioncrest Commercial, Singapore 238467

Cambridge University Press is part of Cambridge University Press & Assessment, a department of the University of Cambridge.

We share the University's mission to contribute to society through the pursuit of education, learning and research at the highest international levels of excellence.

www.cambridge.org
Information on this title: www.cambridge.org/9781009771320

DOI: 10.1017/9781009771337

© Victor Monnin 2026

This publication is in copyright. Subject to statutory exception and to the provisions of relevant collective licensing agreements, no reproduction of any part may take place without the written permission of Cambridge University Press & Assessment.

When citing this work, please include a reference to the DOI 10.1017/9781009771337

First published 2026

A catalogue record for this publication is available from the British Library

ISBN 978-1-009-77132-0 Hardback
ISBN 978-1-009-77137-5 Paperback
ISSN 2517-780X (online)
ISSN 2517-7796 (print)

Cambridge University Press & Assessment has no responsibility for the persistence or accuracy of URLs for external or third-party internet websites referred to in this publication and does not guarantee that any content on such websites is, or will remain, accurate or appropriate.

For EU product safety concerns, contact us at Calle de José Abascal, 56, 1°, 28003 Madrid, Spain, or email eugpsr@cambridge.org

Critical History for Tomorrow's Paleontology

Elements of Paleontology

DOI: 10.1017/9781009771337
First published online: January 2026

Victor Monnin
Archives Henri-Poincaré

Author for correspondence: Victor Monnin, victor.monnin@gmail.com

Abstract: This Element argues for the benefits of integrating the perspectives of a new and growing historiography of paleontology in the training of upcoming paleontologists and in the paleontological community's culture more broadly. Wrestling with the complex legacy of its past, the paleontological community is facing the need to reappreciate its history to address issues of accessibility and equity affecting the field, such as gender gap, parachute science, and specimen repatriation. The ability of the paleontological community to address these issues depends partly on the nature of its engagement with the past in which they find their source. This Element provides a conceptual toolkit to help with the interpretation of the unprecedented position in which the paleontological community finds itself regarding its past. It also introduces historiographical resources and provides some suggestions to foster collaboration between paleontologists and historians.

Keywords: history of paleontology, philosophy of history, historiography, critical history, inclusion, equity

© Victor Monnin 2026

ISBNs: 9781009771320 (HB), 9781009771375 (PB), 9781009771337 (OC)
ISSNs: 2517-780X (online), 2517-7796 (print)

Contents

1 Introduction: The Presence of the Past　　　　　　　　　　1

2 Three Species of History　　　　　　　　　　　　　　　　10

3 Textbook History of Paleontology　　　　　　　　　　　　19

4 A New Historiography of Paleontology　　　　　　　　　　27

5 Cultivating Critical History　　　　　　　　　　　　　　　33

6 Conclusion: Bridging Two Historical Disciplines　　　　　　37

7 Thematic Bibliography　　　　　　　　　　　　　　　　　40

　　References　　　　　　　　　　　　　　　　　　　　　46

1 Introduction: The Presence of the Past

As an historian, my contribution in this Element will be centered on the historiography of paleontology, or the different ways in which paleontologists and historians have been engaging with paleontology's past. The main thesis is as follows: How the history of paleontology is being told and shared has a role to play in making paleontology more accessible[1] and equitable[2]. How any community, including the paleontological community, relates to its past and summons it in the present helps define some of its future. This Element introduces a new and growing historiography of paleontology as a valuable resource for anyone (students, educators, researchers, curators, volunteers, community leaders, etc.) seeking to make the study of the Earth's deep past benefits to and from a diversity of people worthy of the diversity of life it brights to light.

1.1 Appreciating the Presence of the Past

Paleontologists are in a privileged position to appreciate the presence of the past. They are, as Norman MacLeod once put it, "no strangers to an obsession with the past" (MacLeod 2006). The nature of their work brings to light the simple fact that the past is never truly gone. Past life lingers in the present, leaving behind traces of things that happened and remains of beings that were alive long ago. Paleontologists deal with a large spectrum of traces and remains, from faint ichnofossils[3] to incredible Lagerstätte.[4] Advances in technology and research methods have been providing ways to retrieve clues believed to be forever erased and to get around the gaps in the fossil record. From a paleontological perspective, the past is also present *within* the forms and structures of today's organisms and environments. This constitutive place of the past within the present is what provides grounds for the analogies paleontologists draw between the extinct and the existent and for the relevance of paleontological knowledge to understand our world (Halliday 2022, p. xv–xx).

Since the early beginnings of the field, paleontologists have expressed their deep appreciation for the presence of the Earth's past. They have shared the unique and fulfilling experience of being able to "see" the past in the present (Turner 2019). Recounting his climbing of a mountain in the

---

[1] In this Element, accessibility refers to the actual ability for someone to access paleontological education, training, careers, and other opportunities depending on their physical and cognitive abilities, gender, origins, economic means, and cultural background.

[2] In this Element, equity is the attempt to address the different needs and to acknowledge the contributions of distinct groups and individuals in a fair manner that does not benefit some while disadvantaging others.

[3] Ichnofossils consist in fossilized traces revealing the activity of extinct organisms.

[4] A Lagerstätte is a sedimentary deposit preserving fossil remains in excellent conditions and offering an unusual amount of paleontological information.

Andes, Charles Darwin, then a young geologist embarked on the Beagle expedition, described how the landscape surrounding him was rendered even more sublime by finding fossil shells on the most elevated ridge:

> The atmosphere resplendently clear; the sky an intense blue; the profound valley; the wild broken forms; the heaps of ruins, piled up during the lapse of ages; the bright-coloured rocks, contrasted with the quiet mountains of snow; all these together produced a scene I never could have figured to my imagination. Neither plant nor bird, excepting a few condors wheeling around the higher pinnacles, distracted the attention from the inanimate mass. I felt glad I was alone: it was like watching a thunderstorm, or hearing a chorus of the Messiah in full orchestra. (Darwin 1840, p. 394)

About a decade later, Gideon Mantell concluded his geological guide to the Isle of Wight by superimposing the deep past onto the present in an exercise of stereoscopic description:

> At the present time, the deposits containing the remains of the mammoth and other extinct mammalia, are the sites of towns and villages, and support busy communities of the human race; the Huntsman courses, and the Shepherd tends his flocks on the elevated masses of the bottom of the ancient chalk ocean – the Farmer reaps his harvests upon the cultivated soil of the delta of Iguanodon – and the Architect obtains from beneath the petrified forest, the materials with which to construct his temples and his palaces. (Mantell 1847, pp. 412–413)

Experiencing the presence of the past has been and remains a leitmotiv in the popular writings of many paleontologists. In *Dinosaur Heresies*, Robert T. Bakker shared with his readers some of his experience working in the field in the early morning. In the mind's eye of the paleontologist, the fossils still trapped in the rock seem to find life again:

> From where I sit on the quarry's rim I can see the dinosaur's great trochanters, the attachment site of the immense hip muscles, and the bone surface pitted and rough where tendons and ligaments were anchored to the femur. A hundred thousand millennia ago, those tendons and muscles were full of dinosaur blood coursing through capillary beds, bringing oxygen to the cells that powered the stride of this ten-ton giant. Muscles pulsed in cycles of contraction and release, and the hind limb, fully twelve feet long from hip to toenails, swung through its stroke covering six feet with every pace. (Bakker 1986, p. 31)

Sharing the details of his expedition in Mongolia at the beginning of the 1990s, Philippe Taquet described in similar terms his encounter with fossil remains:

> For us today, in the twilight of the end of a beautiful day, in the hollow of this canyon in the Nemegt Valley, in front of the parched remains of this superb

Figure 1 *Bill Turnbull's Eocene* by Adrienne Stroup. 2016. Reproduced with the generous permission of the artist.

tarbosaur with its skin still preserved, truth became stranger than fiction, and the shadow of the past was truly present. (Taquet 1998, p. 146)

One recent and compelling expression of this appreciation for the presence of the past is a picture made in 2016 by Chicago Field Museum collection assistant and paleoartist Adrienne Stroup, *Bill Turnbull's Eocene* (Figure 1). The top-half of the picture, inspired by black-and-white photographs from the 1950s (Stroup 2017), shows paleontologist William Turnbull working in southern Wyoming. The landscape is barren and, apart from the lone protagonist, graced with very few signs of life. In contrast, the colorful bottom-half of the picture shows a river running through a luxuriant landscape inhabited by animals from the Washakie Formation. The two scenes are cleverly connected by the posture of Turnbull, who, bending over his sieve in search of fossils, appears to be peaking at the scene from the geological past unfolding beneath him.

The fading of the sky's blue in the black-and-white sands completes the illusion. Strangely enough, the 1950s seem more remote than the Eocene period.

Stroup's visual tribute to Turnbull points to the fact that the paleontological community also has a tradition of appreciating its own past. No scientific field can thrive as such without its current members acknowledging the work of their predecessors and, by doing so, forging connections that unites them through time and space in one research community. Paleontology is a "science of the archive" (Daston 2012), not only because it deals with fossils, the "archives" of life, but also because it relies on the archived work of predecessors. Studying the depths of geological time, just like studying the immensity of the universe, requires collaboration across many generations. When paleontologists study fossil specimens stored in museum or university collections, they deal simultaneously with natural "archives" and human archives, remains of past life informed by the work of predecessors. It is the intergenerational sum of this work, which sometimes adds its own puzzling history (Stimpson et al. 2024), that allows for paleontological research to generate new insights. One example of such intergenerational archival work is the global cataloguing of foraminifera, first spearheaded by micropaleontologists Brooks F. Ellis and Angelina R. Messina. Funded by the United States Work Progress Administration in the midst of the Great Depression of the 1930s (Croneis 1942), the collection and classification of fossil foraminifera turned into an endeavor spanning across borders and decades, way beyond the lifespan of its original editors, who, by the time of Ellis' passing, had published sixty-nine volumes. Today, more content continues to be added to the catalogue on a yearly basis, making it an almost century-long archival project.

The paleontological community has a long tradition of acknowledging the contributions of its past members and of writing stories of foundation and growth for its science. In 1899, Karl Alfred von Zittel published the first authoritative history of geology and paleontology, emphasizing the international dimension of these two fields: "The questions of the highest import in Geology and Palaeontology are in no way affected by political frontiers" (Zittel 1901, p. v).

In 1931, Henry F. Osborn, then Honorary Curator of vertebrate paleontology at the American Museum of Natural History (AMNH), published *Cope: Master Naturalist*. Dedicated to the life and accomplishments of Osborn's late mentor, Edward D. Cope, this copious volume laid out a narrative of the early development of paleontology in the United States beginning with an age of "pioneers," including Thomas Jefferson and Edward Hitchcock, leading on to an age of "founders," including Joseph Leidy, Othniel C. Marsh and Cope himself. While Osborn was eager to preserve and celebrate the achievements of these men, he mostly hoped that his book would bring "posthumous justice" (Cope 1931, p. vii) to his mentor. Osborn's biography of Cope and early history of American paleontology provided a storyline which greatly influenced the way the paleontological community (particularly its American contingent) tells its own story.

In the 1940s, George G. Simpson, then Associate Curator of vertebrate paleontology at the AMNH, followed in Osborn's footsteps and published a couple of articles elucidating the beginnings of vertebrate paleontology in North America (Simpson 1942, 1943). In 1960, J. Marvin Weller presented in the *Journal of Paleontology* a four-part periodization of the "development of paleontology" from Aristotle to Simpson (Weller 1960). In the 1980s, Henry N. Andrews published a history of paleobotany, *Fossil Hunters: In Search of Ancient Plants* (Andrews 1980), Eric Buffetaut composed *A Short History of Palaeontology* (Buffetaut 1987), and Stephen J. Gould offered, among many other writings, an essay on historical conceptions of geological time, *Time's Arrow, Time's Cycle* (Gould 1987). Beginning in the 2000s, Philippe Taquet published the first two volumes of what was intended to be a three-part biography of Georges Cuvier (Taquet 2006, 2019). In 2018, Adrian Lister retraced the history of Darwin's fossil collecting during his voyage on the Beagle (Lister 2018).

It is clear, then, that since its beginnings, the paleontological community has shown not only a strong appreciation for the presence of the Earth's past, as one would expect, but also an active engagement with the presence of its own past. With an ongoing tradition of historical writing, the paleontological community has, over the years, worked on interpreting and recovering some of its past.

1.2 *Re*-appreciating the Presence of the Past

Recently, a growing number of initiatives from within the paleontological community has begun to challenge this tradition. For their authors, the past of the paleontological community should be *re*-appreciated. This marks an unprecedented turn in the history of the paleontological community's engagement with its own past. Much has been written about the history of paleontology, but shouldn't we look at it again? Such is the question that has been gaining in strength in recent years. A question which raises a series of others: Who is the "we" that has been writing the history of paleontology and its community? Whom did that "we" include and exclude? Whom did it remember and forget? How does the "we" of today compare with the "we" of yesterday? The present paleontological community is knocking on the door of its own past with a renewed set of interests, less concerned with pioneering ages, stories of great discoveries, and spurts of progress. Instead, the reappreciation of the paleontological community's past is concerned with persistent inequities and problematic practices that might impede on paleontological research and its social implications. It is not a search for origins but a search for perspectives on present issues. While the tradition of appreciating the past had been about building a community, with its shared stories and founding figures, the growing efforts to reappreciate the past are about reforming and reimagining that community.

During the spring and summer of 2014, paleontologist Ellen Currano and filmmaker Lexi Jamison Marsh asked women working in paleontology to be photographed in the field or in their laboratories wearing fake beards. The result is a series of provocative, often humorous, black-and-white portraits (Figure 2) parodying the iconic pictures of rugged "fossil hunters" (Figure 3) that have come to mostly define the visual and popular identity of paleontology (Panciroli 2017). As an artistic commentary on the history of gender inequities within paleontology, the "Bearded Lady Project" (Marsh and Currano 2020) brings to light the necessity for the paleontological community to engage creatively with its own past, both through words and images, if it is to benefit from and to all interested individuals. This project is but one contribution within a broader and

Figure 2 Leckie Lab: University of Massachusetts Amherst: Adriane R. Lam, PhD candidate in micropaleontology; Raquel Bryant, MS student of micropaleontology; and Serena Dameron, PhD candidate in micropaleontology Photograph by Kelsey Vance. 2015. Courtesy of The Bearded Lady Project.

Figure 3 Othniel C. Marsh with his assistants. 1871. Image obtained from Wikimedia Commons.

growing conversation on the inclusion of people who have historically been either excluded, marginalized, or neglected in paleontology and the geosciences due to their non-conforming gender identities (Olcott and Downen 2020), their neurodiversity (Kingsbury et al. 2020), or their disabilities (Chiarella and Vurro, 2020).

More recently, in 2022, a collective of researchers published in *Nature Ecology & Evolution* the results of a scientometric study showing how historical events and their legacy continue to play a part in the sampling biases that characterize the fossil record. By arguing that "Spatial sampling biases are borne out not only by geological and physical factors influencing the fossil record, but also by pervasive historical and socioeconomic factors" (Raja et al. 2022, p. 150) that perpetuate international asymmetries in paleontological contributions, the authors highlighted the necessity to investigate the past of the paleontological community from a global and economic perspective. The results of the study also made clear that if the paleontological community stretches beyond political frontiers, as Zittel pointed out at the turn of the twentieth century, its history should not be hastily reduced to one of international collaboration and competition. Indeed, the means and opportunities to collaborate or compete have not been shared in equitable measures (Dunne et al. 2025). For the authors of the study, finding ways to measure historical factors affecting access to fossil material is a necessary step toward achieving the inseparable goals of improving our understanding of deep-time biodiversity and fostering global equity in paleontological research.

That same year, Pedro Monarrez and colleagues published in *Paleobiology* a piece hoping "to encourage further discussion about colonialism and racism in geology and paleontology" (Monarrez et al. 2022, p. 181). As with the two previous examples, the past of the paleontological community is approached by the authors as a space that needs to be reappreciated. At the beginning of their article, the authors acknowledge that their reflections and research were significantly motivated by the "racial unrest in the summer of 2020" (Monarrez et al. 2022, p. 174) that followed the murder of George Floyd in Minneapolis; a tragic event which resonated worldwide with other contexts of racial inequities (Klein 2020) and occurred at the beginning of the COVID-19 pandemic which was already casting a harsh light on economic disparities across the world (Stiglitz 2022). Some members of the geosciences community responded to these events by implementing the virtual and comprehensive initiative "Unlearning Racism in Geoscience" (URGE), which offers anti-racism curriculum and resources "to deepen the community's knowledge of the effects of racism on the participation and retention of Black, Brown, and Indigenous people in Geoscience."[5]

If extraneous factors and events have encouraged a renewed engagement with the past through the lens of social issues, these same attempts are also determined by changes affecting the field of paleontology more directly. Among these changes is the growing number of women working in the field. According to the Paleontological Research Institution website, the percentage of women holding a membership in the Society for Vertebrate Paleontology doubled between 1980 and 2017, from 18% to 36%.[6] One effect of this increase over the last forty years is that it has rendered even more striking the stagnant numbers of racial and ethnic diversity over the same period (Bernard and Cooperdock 2018). Additionally, recent debates over specimen repatriation, such as the ones around the holotype specimen of *Ubirajara jubatus*, eventually returned to Brazil in 2023, have seen historical considerations play crucial argumentative functions. Focusing on issues regarding paleontological research conducted in Mexico and Brazil, Juan Carlos Cisneros and colleagues argued that:

> [The] structure of colonial science – derived from the practice of science in the colonies – has given rise to 'scientific colonialism' in the post-colonial world, some of whose extractive practices are sometimes referred to as parachute science, helicopter research, or even parasitic science. Within scientific colonialism, middle- and low-income countries are perceived as suppliers of data and specimens for the high-income ones, the contributions

---

[5] https://urgeoscience.org/development/. Accessed on August 5, 2025.
[6] https://www.museumoftheearth.org/daring-to-dig/women-at-the-forefront. Accessed on August 4, 2025.

of local collaborators are devalued or omitted, and the legal frameworks in lower income countries are trivialized or even ignored. In turn, colonialist nations, owe their wealth to these colonial practices that have existed for centuries, allowing them to accumulate knowledge, power and financial resources. These extractive practices persist in the field of palaeontology to this day. (Cisneros et al. 2022, p. 2)

More broadly, the movement to decolonize natural history collections and develop renewed narratives around them to educate the public (Das and Lowe 2018, Armstrong and Sharp 2024, Hide 2024, Hurst et al. 2024) has also been encouraging the re-appreciation of paleontology's past. In 2018, a team of researchers from the Museum für Naturkunde in Berlin revisited the history of the specimens of the *Brachiosaurus Brancai* found at the beginning of the twentieth century at Tendaguru Hill, in current Tanzania, questioning the inherited narrative of its discovery:

In additions to deconstructing the narrative, we also seek to recontextualize it – that is, to place the expedition and its finds within their broader historical, sociopolitical and musicological contexts. ...The Tendaguru specimens uniquely embody the complex ties between colonialism, natural history and museums, and we use them to examine the conditions that permitted these and other fossils to be claimed, extracted, removed from their countries of origin and appropriated. ...Given this history, the Tendaguru fossils raise the urgent question of who 'nature' really belongs to. (Heumann et al. 2024, p. 4)

For the paleontological community, these new forays into the past make its history both harder to recognize and more urgent to understand. It is not surprising that these reappreciations of the past are sometimes met with skepticism or resistance. Some will want to defend the legacy of the "pioneers" and "fathers," whose scientific accomplishments, they fear, might be obscured by anachronistic moral judgments. But the value of these efforts to reappreciate the past of paleontology does not lie in an attempt to separate "heroes" from "villains." Instead, their value resides in the desire to engage with the past in ways that can help the paleontological community reconsider its inherited practices, culture, and institutions to improve its scientific production and social relevance. To build a sustainable paleontological community for tomorrow, a new, shared approach to the field's past is needed.

1.3 Objectives and Plan

The first objective of this Element is to provide a conceptual toolkit to help with the interpretation of the unprecedented position in which the paleontological community finds itself regarding its own past. The second objective is to present historiographical resources and provide suggestions to assist the members of the paleontological community in engaging with the past of their field in the most beneficial way.

Section 2 details three different ways in which a community can engage with its past, following a distinction drawn by the philosopher Friedrich Nietzsche. His distinction of three "species" of history (monumental, antiquarian, and critical) is mobilized in section 3 to assess how the paleontological community has traditionally been engaging with its past and what the limitations of this kind of engagement are. Section 4 introduces recent perspectives in the historiography of paleontology and explains how they could benefit the paleontological community in cultivating a renewed engagement with its past. Section 5 provides some suggestions on how to cultivate this new engagement with the past at the instruction, publication, and research levels. Standing as a conclusion, Section 6 reflects on the similarities that paleontology and the history of paleontology share as historical disciplines, and how these similarities should encourage cooperation for the mutual benefits of members of the paleontological community and historians. Section 7 consists in a thematic bibliography designed to facilitate the exploration of recent perspectives in the historiography of paleontology.

2 Three Species of History

2.1 An Historical Reckoning

It appears that the paleontological community is currently living through an historical reckoning. In light of the examples discussed in the previous section, the sources of this reckoning are numerous and complex, coming both from within and beyond the paleontological community. But it is certainly for the first time, in the two centuries of its existence, that this community is confronted with the emergence, within its own ranks, of historical discourses and commentaries that situate the field within broader ethical, social, economic, and global problematics. Instead of contributing to a more traditional narrative of progress in knowledge and methods, these discourses and commentaries point at inequities which, they argue, have limited access to the field in the past and which legacies continue to do so at the present time. The emergence of this new approach raises the question of what the paleontological community should do with its own past. Should it care about the broader historical context of the field's developments? Should it leave such matters to historians? Should it limit itself to a version of its history focusing strictly on scientific matters? How much credit should it give to these new historical discourses? And if they deserve any, what good can they serve? The paleontological community is living through an historical reckoning because it faces, more urgently and explicitly than ever before, the need to choose the way in which it will engage with its own past.

Such interrogations are inevitably sources of tensions. Some might welcome the rise of new approaches to the field's past as a much needed response to present challenges, while others might interpret it as an unwarranted attack on a past that had usually been a source of pride or, at the very least, a consensual object of intellectual curiosity. On both ends of the spectrum, there will be reasons to feel strongly about how the field's past should and shouldn't be approached. The reason behind such disagreement, I argue, lies in the variety of ways in which a same community can engage with its own past and which one(s) its members have learned to practice and value. To address this complex and delicate situation, it is therefore necessary to be able to distinguish between different ways of engaging with the past. Drawing clear distinctions should facilitate a productive conversation over what the paleontological community should do with its own past, how it has dealt with it until now, and how it might need to deal with it tomorrow.

To draw such distinctions, I propose to rely on an influential piece of philosophy of history, "On the Uses and Disadvantages of History for Life," an essay published by German philosopher Friedrich Nietzsche in 1874. In this essay, Nietzsche distinguishes between three "species" of history, or ways in which a given community can approach its own past. The philosopher named these three species, *monumental, antiquarian*, and *critical*. Acknowledging the irreducible need for history in the life of any human community, Nietzsche tackled the problem of evaluating when "the study of history is something salutary and fruitful for the future" (Nietzsche 1997, p. 67) and when it ceases to be so.

To each species of history (monumental, antiquarian, and critical), the philosopher recognizes advantages and disadvantages. This chapter offers a working characterization of each species and shows how they can be used as conceptual tools to analyze the different ways in which the paleontological community has been engaging with its past. The aim of this section is not to comment on Nietzsche's philosophy of history,[7] but rather to show how his distinction of three species of history can readily be mobilized to help clarify the challenge that the paleontological community is now facing regarding its own past. The following sections will discuss, in order, monumental history,

---

[7] The scholarship on Nietzsche's philosophy of history is extensive and discusses the philosopher's evolving thoughts on history and its different forms. "On the Uses and Disadvantages of History for Life" should certainly not be considered as Nietzsche's last words on the subject (Brobjer 2004).

antiquarian history, and critical history, illustrating each of them with examples related to the paleontological community.

2.2 Monumental History

Monumental history, as for the other two species of history, answers in a specific way the necessity that a given community has to engage with its past. As its name indicates, practicing monumental history consists in searching the past for "monuments," landmarks of great accomplishments performed in the face of adversity, which may serve as sources of inspiration in the present. Looking at the past through such a lens, one "learns from it that the greatness that once existed was in any event once *possible* and may thus be possible again" (69). The value of monumental history lies in its ability to turn the past into a collection of examples that remind us that great things have been and can still be achieved despite doubts and difficulties.

Like any other community of science workers, the paleontological community has been engaging in monumental history to find, in its past, examples of great figures and accomplishments to emulate. In fashioning a collection of bright examples, a scientific community can simultaneously find pride in what has been accomplished by its predecessors and hope in what can be achieved by future generations. The monumental investment in the past by the paleontological community can be observed in a number of written tributes, events dedicated to the memory of past members, and actual monuments, such as statues of important figures in the field (Figure 4). All of these serve the important purpose of remembering "models, teachers, comforters" (67) to look up to when tackling the great tasks of the present. A most compelling example of the paleontological community cultivating a form of monumental history can be found, as it is the case in all other scientific fields, in the naming of prizes and awards after prominent contributors. Designed to recognize the accomplishments of younger members, such as the Alfred Sherwood Romer Prize from the Society of Vertebrate Paleontology, or more seasoned members, such as the Raymond C. Moore Paleontology Medal from the Society of Sedimentary Geology, these honors are also meant to actualize, in the form of periodic rituals, a connection between the great achievements of the past and those of the present. Year after year, the growing list of awardees constitutes its own memory of great accomplishments to look up to.

Members of the paleontological community continue to look for and elevate figures from the past which they believe could inspire their community to surmount present challenges in the field. In recent years, for example, a number of initiatives have brought to light the accomplishments of pioneering

Figure 4 Statue of Georges Cuvier by David d'Angers. 1835. Montbéliard, France. Image obtained from Wikimedia Commons.

women in paleontology in the hope of (a) providing paleo-enthusiastic girls and women with role models and (b) encouraging future efforts toward gender equity in the field. As the webpage of the 2020 exhibition *Daring to Dig: Women in American Paleontology* explains: "Despite [. . .] barriers, pioneering women found ways to surpass stereotypes, circumvent obstacles, and forge new paths in their efforts to contribute to the science of paleontology." Presenting a gallery of women who contributed to the field from the seventeenth century to the present day, this exhibition could be interpreted as an exercise in monumental history, as far as its ambition to elevate new figures from the past who found ways to weather adversities is concerned.[8] In a similar effort, although taking quite a different form, a statue commemorating the British fossil collector and paleontologist Mary Anning (1799–1847) was inaugurated in Lyme Regis in May 2022 (Figure 5). This inauguration was the culmination of a campaign,

---

[8] *Daring to Dig* could also be characterized as an initiative critically engaging with the past. As we explain in the last part of this section, the three species of history do not always correspond to clear-cut ways of engaging with the past. Sometimes, characteristics of different species of history can be observed in the same initiative or publication.

Figure 5 Bronze sculpture of Mary Anning by Denis Dutton. 2022. Lyme Regis, Dorset, England. Image obtained from Wikimedia Commons.

"Mary Anning Rocks," launched by a local schoolgirl and her mother four years prior. Such initiatives show that, despite what the adjective may suggest, monumental history is not a static way of engaging with the past. It is a continuous search for figures from the past to erect as monuments for the benefit of the present. New monuments arise, others crumble, while some endure, depending on what sort of inspiration members of the paleontological community believe they most need from the past.

Despite all its benefits for the present and future of a community, some disadvantages are also associated with monumental history. The elevation of any figure from the past to the status of example runs the risk of idealizing that figure by making abstraction of details and circumstances which would prevent it from serving as an equivocal source of inspiration. Indeed, "as long as the past has to be described as worthy of imitation, as imitable and possible for a second time, it of course incurs the danger of becoming somewhat distorted, beautified and coming close to free poetic invention" (Nietzsche 1997, p. 70). The risks of monumental history lie in how the "beautified" version of the past might, unseemly, come to (a) impoverish the past by stripping it of its nuances and complexities, and (b) eventually stand for the only past that really matters.

These risks, we will have to keep them in mind when, in Section 3, we will be discussing how the history of paleontology has traditionally been presented to aspiring members of the field in textbooks.

Monumental history is a way of looking at the past in search of inspiring examples to tackle the great tasks of the present with confidence. But if this engagement with the past answers a most important and collective need for role models, it also runs the risk of forgetting "whole segments" of the past (71), leaving it idealized, partial, and misunderstood.

2.3 Antiquarian History

In contrast to monumental history, antiquarian history is not searching the past for examples to emulate in the present, but is searching the present for past things to cherish and pass on to the next generations. Engaging in antiquarian history is "tending with care that which has existed from old" (72–73). Such an engagement with the past is motivated by the need to find and preserve the roots of one's own existence and what made it possible. Its main advantage for the life of individuals and communities lies in the cultivation of a feeling of both belonging to a larger whole and an appreciation for what was inherited from the past. It is akin to "the contentment of the tree in its roots" (74).

In addition to the examples of monumental history discussed earlier in this chapter, members of the paleontological community have also been practicing antiquarian history. Just as turning certain individuals of the past into examples to emulate is important to encourage future accomplishments in the field, caring for what was left behind by preceding members of the field is important to foster a sense of (a) appreciation for the continued existence of the field, and (b) belonging to its intergenerational community. This form of engagement with the past can be witnessed in anniversary events or publications celebrating foundational moments in the history of the field and providing opportunities to appreciate and reflect on the work accomplished since then. In 2024, for example, the paleontological community, especially its members interested in dinosaurs, celebrated the 200-year anniversary of what is considered to be the first valid scientific description and naming of dinosaur remains, published by British geologist and paleontologist William Buckland in the *Transactions of the Geological Society* of 1824 (Buckland 1824). Among the many special publications and events marking the bicentennial of the naming of *Megalosaurus*, the Natural History Museum in London hosted an international conference at which were exhibited the original jaw fragment described by Buckland along with an original copy of his publication (Figure 6). The value of having the original specimen at such an event resides, of course, in it being the

Figure 6 The *Megalosaurus* jaw from the Oxford University Museum of Natural History exhibited at the Natural History Museum in London. Photograph by Emily Keeble 2024. Reproduced with the generous permission of the photographer.

first holotype of a dinosaur, but even more so in the emotional response it might prompt in those who, two centuries later, can "see" in it the origins of their field and their own work. The care taken in preserving and exhibiting such remains from the past is not only justified by their obvious scientific value, but also by their value as roots for an entire community of dinosaur specialists.

Despite these clear benefits, there remain potential dangers associated with an excessive antiquarian approach. While such engagement with the past provides useful opportunities for a community to cultivate meaningful connections with its past, its feeling of appreciation for what came before might run the risk of becoming too acute. As is the case with an excessive monumental approach, an unchecked antiquarian approach limits the ability to fully understand the past. Focused on the care and respect for roots, "[t]he antiquarian sense of a man, a community, a whole people, always possesses an extremely restricted field of vision" (Nietzsche 1997, p. 74). While monumental history can lead from inspiration to idealization, antiquarian history can lead from appreciation to reduction. In both cases, the challenge lies in the ability of a community to remain aware of the many ways in which it engages with its past.

2.4 Critical History

A critical outlook on the past does not consist in a search for inspiring examples, nor in a search and care for roots. It consists in an examination of parts or aspects of the past seen as particularly unjust and which legacies are still affecting the present. Instead of seeking inspiration from the past as monumental history does, or seeking a sense of appreciation and belonging from the past as antiquarian history does, critical history "seeks deliverance" (67) from the unjust things that have been inherited from the past. The need for critical history arises when forgetting certain past injustices is no longer an option and there is a need "to be clear as to how unjust" (76) something inherited from the past is.

This sort of engagement with the past, the paleontological community has recently seen it emerge within its own ranks. The "Bearded Lady Project", cited in the introduction, is one example of such a critical engagement with the past, addressing, in this specific instance, the legacy of gender discrimination in paleontology. Other injustices inherited from the past have also begun to be addressed more explicitly, prompting the recovery of stories neglected until now. Since the second half of the 2010s, a collective of researchers has been working on unearthing the history of indigenous paleontology in South Africa. Blending methodologies from geomythology,[9] archeology, paleontology, and archival research, this initiative is helping evaluate how indigenous people's cultural interpretations of fossils might have contributed to modern re-discoveries of fossil sites and could continue to do so in the future (Benoit 2018, Benoit et al. 2024, and Helm et al. 2018). Such an endeavor provides historical evidence to help develop a more equitable culture of recognition and credit in paleontology (Valenzuela-Toro et al. 2025). In 2023, Hannah Kempf and colleagues published a historical review of the United States' policies governing paleontological research on Native American lands (Kempf et al. 2023). This publication aimed at bringing the attention of the paleontological community, not only toward an "unfortunate history of fossil dispossession from Native American lands" (200), particularly during the mid-nineteenth century, but also toward present "policy gaps and ambiguities surrounding fossil collection on Native American lands" (194) which perpetuate, granted in different ways than a century and a half ago, this history of dispossession. In this case, critical history entails the gathering of information and evidence necessary to recount past injustices, evaluate their repercussions on the present, and suggest future ways of remedying them.

While critical history might allow a community to recognize and address old injustices which effects and legacies it does no longer tolerate, this form of engagement with the past "is always a dangerous process" (Nietzsche 1997, p.76), as those criticizing the injustices of the past still remain themselves "the outcome of earlier generations, [...] of their aberrations, passions, and errors"

---

[9] Coined by Dorothy Vitaliano, the term "geomythology" referred originally to the project of explaining "certain specific myths and legends in terms of actual geologic events that may have been witnessed by various groups of people" (Vitaliano 1973, p. 1). Now, the term also refers to the study of cultural interpretations of fossil remains and of geological features and events.

(76). The burden of past injustices is not easily removed, and this persistence might make any efforts to address them appear futile. Another potential risk is that of adopting a critical outlook on the past without the clearly recognized need of addressing any specific injustices in the present. Since "every past [...] is worthy to be condemned – for that is the nature of human things: human violence and weakness have always played a mighty role in them" (76), criticizing the past for its own sake could lead one down the path of raising an interminable list of retrospective grievances and condemnations without any clear benefits for the present.

2.5 A Conceptual Toolkit

This distinction between three species of history (monumental, antiquarian, and critical) can be used as a conceptual toolkit to clarify the nature of the historical reckoning the paleontological community is currently facing. Mobilizing these three species should help identify, evaluate, and compare the different ways it has been engaging with its past. While the following sections propose an interpretation of this historical reckoning and suggest steps to approach it, the main objective in laying out the distinction between three species of history is to provide conceptual tools for the benefit of a constructive conversation about what the paleontological community has been and should be doing with its past.

Before proposing an interpretation of how the paleontological community has traditionally been engaging with its past and transmitting it to its upcoming members, it is important to point out that the three species of history, like any other conceptual distinctions, do not neatly correspond to clear-cut observable ways of engaging with the past. Individuals and communities entertain complex, and often contradictory, relationships with the past. Characteristics of monumental, antiquarian, and critical histories might be recognizable in the same initiative or publication, and, as we saw throughout this section, the same community engages with the past in multiple ways simultaneously. Monumental, antiquarian, and critical histories are only the tools this Element proposes to use to analyze, evaluate, and compare the ways in which the paleontological community has been engaging with its past. They do not stand for the ways themselves.

It is also important to underscore that these three species of history, as conceptualized by Nietzsche, are part of a well-established canon of Western philosophy. The decision to mobilize them in this Element does not presume their superiority over other ways of thinking about the past. Arguably, the development of an always more inclusive paleontological community calls for the recognition and integration of diverse worldviews and relationships to human and non-human pasts, in particular from non-Western and Indigenous cultures (Howe and Rieppel 2024, Hurst et al. 2024). On one hand, the decision to rely on this threefold distinction is a reflection of my European cultural background and academic training. On the other hand, this distinction can easily be mobilized to begin disentangling some ambiguities inherent to the value-laden and polysemic notion of history, facilitating therefore a constructive conversation over the paleontological community's engagement with its own past and why it matters.

3 Textbook History of Paleontology

There are many places and times for members of a given scientific community to learn about the history of their field and to develop a certain picture of its past. Surely, this part of any member's education can be experienced as a continuous process by those who, either because of their research agenda or their personal curiosity, actively engage with the work of some of their predecessors. But among the variety of ways through which individual members of the same community can come to learn about the past of their field, textbooks constitute a key locus for the articulation, transmission, and consolidation of a shared narrative. For this reason, taking a closer look at how the history of paleontology has been told in introductory textbooks should provide valuable insights on how the paleontological community has traditionally been intending to teach its upcoming members to approach the field's past.

Focusing solely on textbooks has obvious limits, since instructors might assign additional readings and other contents, prepare materials, and give lectures on historical topics beyond what is covered in the textbooks they use as references, or avoid textbook historical exposés altogether. Textbooks are also not accessible to all students equally and not all vocational paleontologists took formal coursework in paleontology or related fields. A much broader study, including a survey of instructors, would be necessary to draw more solid conclusions on at least the current state of the teaching and transmission of the field's history. Nevertheless, textbooks, as cornerstone publications dedicated to the training of aspiring members of the field, can be approached as artifacts recording over time the kind of history and the place generally attributed to it in such training. They constitute a relatively homogeneous group of traces that can be compared with each other to begin reconstructing the evolution of the teaching of paleontology's history. Although this method might not bring the most precise conclusions, it suits the aim of this section to help clarify the complex position in which the paleontological community finds itself regarding its relationship to its own past.

After reviewing a small selection of textbooks, this section mobilizes the three species of history to (1) characterize the kind of engagement with the field's past that has traditionally been transmitted to students of paleontology, and (2) evaluate the advantages and disadvantages of this kind of engagement.

3.1 A Short Review of Textbook Histories

This section analyzes a small selection of textbooks to provide the reader with a general idea of how the history of paleontology has traditionally been presented to students. All the textbooks selected have at least one segment dedicated to the history of the field, but, of course, not all textbooks reserve distinct paragraphs to discuss this topic. This fact alone is significant, for even the decision to not include a discussion of the past constitutes a way to engage with it and might tell us something about a more general approach to the past shared within the authors' community.[10] But rather than speculating on the absence of historical considerations, this section describes how the history has been told when it has been explicitly covered in textbooks.

Published in 1955, Edwin H. Colbert's *Evolution of the Vertebrates* summarizes the history of paleontology in one paragraph:[11]

> To the peoples of the classical civilizations the nature of fossils was hardly realized, and it was not until the days of the Renaissance that a true understanding of fossils was reached by (among others) that great and accomplished Florentine, Leonardo da Vinci. The scientific study of fossils, however, is barely more than a century and a half old, whereas the modern evolutionary interpretation of the fossil record really begins with the work of Charles Darwin, as crystallized in his epochal book *The Origin of Species*. In spite of the relative youth of paleontology as a science, an impressive amount of fossil material has been gathered together and studied by paleontologists all over the world. (Colbert 1955, p. 2)

The speculations of ancient philosophers over the nature of fossils, the genius insights of Renaissance scholars like Leonardo da Vinci, and the revolution prompted by the publication of Darwin's *Origin of Species* are among the main episodes in the history of paleontology that one usually encounters in textbooks. In fact, these episodes can already be found in anterior publications, such as in Marcellin Boule and Jean Piveteau's *Fossils: Elements of Paleontology* (Boule and Piveteau 1935, p. 33–46) or in Charles Schuchert and Carl Dunbar's *Historical Geology* (Schuchert and Dunbar 1941, p. 20–22).

---

[10] "To be a member of any human community is to situate oneself with regard to one's (its) past, if only by rejecting it" (Hobsbawm 1972, p. 3).

[11] Despite the little space attributed to historical considerations in this textbook, Colbert wrote extensively on the history of paleontology (Colbert 1968).

At the beginning of the 1960s, William L. Stokes dedicated a short section to the history of paleontology in his *Essentials of Earth History: An Introduction to Historical Geology* (1960). The author cites the names of Aristotle, Da Vinci, Johann Scheuchzer, and William Smith in a story beginning with the mystical and magical interpretations of fossils by "many primitive and prehistoric peoples" (Stokes 1960, p. 79) and ending with "Smith's methods of correlation" (81).[12] At the end of the same decade, David L. Clark wrote a three-page history of paleontology for the introductory chapter of *Fossils, Paleontology and Evolution* (1968). Clark's historical section references a dozen philosophers and scientists, from Herodotus and Aristotle to Cuvier and Darwin, and frames the story around the resolution of the problem, "are fossils organic or inorganic?" (Clark 1968, p. 2), and how, once this problem had been solved, paleontology was able to consolidate into a science contributing to the study of the Earth's history and evolution.

Fifteen years later, Bernhard Ziegler provided an even richer preface on the "scope and development of palaeontology" in his *Introduction to Palaeobiology* (1983). The story follows the same storyline as Colbert's and Clark's, beginning with the ancient Greeks, noting the "exceptions" (Ziegler 1983, p. 10) of Da Vinci and a few other Renaissance scholars, to finally end with the discipline-defining work of the likes of Cuvier and Smith. Although similar in that respect to previous textbook accounts, Ziegler's three-page preface cites more than forty names, which beyond Herodotus, Aristotle, da Vinci, Cuvier, and Darwin, also includes figures like Bernard Palissy, Benoît de Maillet. James Sowerby, and Othniel C. Marsh. As an addendum to the main narrative, Ziegler dedicates short paragraphs to name some important contributors in the history of micropaleontology, invertebrate paleontology, and paleobotany. The author lists Zittel's *History of Geology and Palaeontology* (1901) and Wilfred N. Edwards' *The Early History of Palaeontology* (1967) as its two historiographical references. Published in the same decade, *Paleontology: The Record of Life* (1989), by Colin Stearn and Robert Carroll, devotes the equivalent of one page to the history of paleontology. The authors evoke the "early speculations" (Stearn and Carroll 1989, p. 5) of Greek philosophers, including Xanthos of Sardis and Aristotle, before mentioning the debates on the origins of "figured stones" during the Middle Ages and the Renaissance, to finally explain how by the beginning of the nineteenth century, the origins of fossils as remains of extinct organisms had been established. The section is illustrated by a plate

---

[12] By the fourth edition of his textbook, Stokes had expanded the historical section with a few paragraphs discussing the contributions of figures like Lamarck, Cuvier, Brongniart, and Darwin (Stokes 1982, p. 102-105).

of figured stones by eighteenth-century Swiss naturalist Karl Land. In contrast to Ziegler's history, this one is almost completely devoid of names. The author cites *The Birth and Development of the Geological Sciences* (1938) by Frank D. Adams as its historiographical reference.

By the end of the 1990s, the textbook history does not appear to have changed significantly. In *Bringing Fossils to Life* (1998), Donald Prothero delivers, in an illustrated, four-page "What is a fossil?" section, a narrative similar to its predecessors in both structure and cast. Dedicating a paragraph to Da Vinci, the author also discusses extensively the work of Nicholaus Steno. Notably, Prothero mentions, although in passing, how the work of Smith on index fossils had been prompted by engineering projects characteristic of the Industrial Revolution. This detail constitutes a rare hint at how broader historical factors might have been a part of the history of paleontology. For further reading, the author suggests Martin Rudwick's *The Meaning of Fossils* (1972).

The first edition of *Introduction to Paleobiology and the Fossil Record* (2009) by Michael Benton and David Harper offers an account of the historical "steps to understanding" of fossils (Benton and Harper 2009, p. 9). The narrative's cast is comparable to the ones described earlier in this section and includes some of the same remarkable episodes. For example, it cites Xenophanes' and Herodotus's "early speculations" (9) and Da Vinci has a short paragraph dedicated to his interpretation of fossil shells observed in the Italian mountains.[13] The section is richly illustrated with, among others, the famous plate showing a shark head from Steno's work on glossopetrae and an illustration by Edouard Riou from Louis Figuier's popular science bestseller *The World Before the Deluge* (1863). In addition to this section, the authors set aside a few paragraphs for some historical comments on "Paleontology and the history of images," from the first drawing reconstructions of extinct animals to *Jurassic Park* (1993) and *Walking with Dinosaurs* (1999). The second chapter offers also a history of biostratigraphy featuring figures like Smith and Roderick Murchison. As in Prothero's textbook, the authors allude to a broader historical context when pointing out that "one of the first systems to be named was the Carboniferous ("coal-bearing"), a unit of rock that early industrialists were keen to identify!" (27–32). Among the reading suggestions related to the history of paleontology, the authors list Rudwick's *The Meaning of Fossils* and *Scenes from Deep Time* (1992) as well as Buffetaut's *A Short History of Vertebrate Palaeontology* (1987). The second edition of Benton and Harper's textbook

---

[13] In the following chapter, the authors include a discussion of "Leonardo's legacy" in geology and how it preceded Steno's laws of the superposition of strata.

from 2020, apart from additions to the "Paleontology and the history of images" section, does not include noticeable changes to the historical account.

A few observations can be drawn from this short review. First, the history of paleontology presented to students in textbooks has not experienced any fundamental changes in, at least, the last seventy years. Despite some differences between the historical accounts discussed in this section, each constitutes a version of a same narrative telling the story of how fossils came to be understood "as fossils," to quote Benton and Harper's expression. Second, the main cast of the story remains notably fixed: Herodotus, Aristotle, Da Vinci, Steno, Cuvier, Smith, and Darwin being among the most systematically cited. Third, apart from a few hints in versions like Prothero's, and Benton and Harper's, the story unfolds with little to no reference to a broader historical context.

Mobilizing our conceptual toolkit to characterize the kind of engagement with the past which appears to have been prevalent in textbooks should allow for an evaluation of the advantages and disadvantages that the paleontological community, especially its upcoming members, may experience from the transmission of such a narrative. The traditional textbook narrative mostly displays characteristics of both monumental and antiquarian histories. It does not only present a gallery of great philosophers and scientists who contributed to solve the "riddle" of fossil remains, but also situates the deepest roots of paleontology as far back as ancient Greece. As the previous section made clear, both monumental and antiquarian histories have advantages and disadvantages for the community who engages in them. The next two sections identify them respectively.

3.2 Advantages of the Traditional Textbook History

The traditional textbook history of paleontology has obvious advantages for the training of prospective members of the paleontological community. As a monumental history, it provides a shared and recognizable set of important past contributors to the field. As an antiquarian history, it also provides an origin story spanning multiple centuries. Both help, as philosopher and historian of science Thomas Kuhn pointed out, in building a sense of community: "Characteristically, textbooks of science contain just a bit of history, either in an introductory chapter or, more often, in scattered references to the great heroes of an earlier age. From such references both students and professionals come to feel like participants in a long-standing historical tradition" (Kuhn 1990, p. 138).

The clear succession of episodes and the relative concision characteristic of the textbook narrative amount to an easily transmissible story, requiring little time and space to reach its goal of fostering a sense of community. Additionally, the lack of references to the broader historical context in which paleontology developed into the scientific field it is today minimizes distraction, which helps in setting the stage for the transmission of the field's most current methods and knowledge, the textbook's raison d'être.

Combining elements of monumental and antiquarian histories, the narrative traditionally found in introductory textbooks serves the paleontological community in providing its members with an inspiring origin story of the field, peopled with prominent figures whose work laid down the foundations of today's science. The narrative serves to remind students that no matter how obvious the nature of fossils may now seem, this certainty is owed to the centuries of intellectual and scientific efforts that preceded them. In this regard, and in addition to its value as a source of inspiration and in community-building, the traditional textbook narrative can also serve to instill a sense of intellectual humility ahead of the work that remains to be accomplished in the field.[14]

3.3 Disadvantages of the Traditional Textbook History

Despite its advantages, the traditional textbook history of paleontology, like any other monumental or antiquarian history, also holds potential disadvantages.

As a monumental history, the textbook history might provide a misleading picture of the kinds of actors that have been contributing to the advancement of paleontological knowledge and methods. Within the limited space assigned to a historical section in a science textbook, choices have to be made and names have to be left out to tell a coherent story. This is arguably the case with any account of the past, even the most comprehensive. The potential disadvantage of the textbook history as a monumental history does not lie in the necessity of selecting, but in the selection traditionally made. The textbook history focuses exclusively on authors who contributed essays, treatises, and scientific publications. While there are many valid reasons to reserve a significant portion of the narrative to this category of actors, focusing exclusively on it leaves unacknowledged a multitude of other actors whose contributions, although not taking a written form, were necessary for the work carried on by those who wrote and published: preparators, artists, workers, miners, guides, informants, and fossil collectors, among others. Again, the spatial constraint of a textbook does not allow for exhaustivity and choices have to be made, but a choice favoring only

---

[14] Benton and Harper insist, for example, on the moral fortitude required to contribute to the growth of paleontological knowledge: "Successful researchers in paleontology, as in any other discipline, need endless patience and stamina" (Benton and Harper 2009, p. 18).

authors runs the risk of delivering an impoverished historical account of the diversity of actors that makes up the paleontological community. The textbook history does not focus exclusively on authors, but on Western men authors. This triple selection obscures even more the interplay of different categories of actors that has been contributing to the advancement of paleontology. Authors stand alone in the traditional narrative, "like a range of human mountain peaks" (Nietzsche 1997, p. 68).

As an antiquarian history, the textbook history focuses on the roots of paleontological knowledge and methods, delivering an account of the history of the field reduced to strict matters of science, such as the nature of fossils or the use of fossils in geology and evolutionary biology. Just as there are good reasons for focusing on authors, there are good reasons for focusing on scientific questions in the confine of a science textbook. After all, the primary goal of a textbook is to introduce readers to the current state of the science. In that case, it seems only appropriate to provide a historical narrative focusing on the science as well. But, similar to the exclusive selection of authors, the exclusive selection of scientific matters provides a narrative dissociating paleontology from any other historical contexts and factors that influenced its own history: the extraction of geological resources, the development of museal institutions, or the establishment of geological surveys in national and colonial territories, among others. The reduction of the history of paleontology to a succession of ideas about fossils and their significance runs the risk of giving the impression that paleontology has been standing and growing aside from "all the rest of the forest around it" (74).

These limitations of the traditional textbook history of paleontology hold unforeseen consequences for the paleontological community today and its upcoming members. As discussed in the previous section, any kind of engagement with the past bears its lot of beneficial and deleterious consequences for the community that practices it. The intended benefits of the traditional textbook history of paleontology come at a price. By limiting itself to authors, it offers an impoverished portrait of the actors that have historically taken part in the advancement of the science. By limiting itself to the succession of ideas, it gives out the impression that this advancement was not itself imbedded in broader contexts. The result is a narrative deprived of human complexity. As historian of science George Sarton once wrote, "such a history becomes quite dehumanized; it is not a true history, but merely the shadow of one" (Sarton 1949, p. 136).

One unforeseen disadvantage of this oversimplification is that the past is not presented as carrying into the present a complex legacy which informs, not only the state of today's paleontological knowledge and methods, but also the current sociology, geography, economy, practices, and institutions of the field. It risks disconnecting present challenges within the paleontological community from their historical component, making them even harder to recognize, understand, and address. At a time that "palaeontology, and the geosciences more broadly, are currently grappling with overdue conversations around the societal and economic impact of their research practices" (Dunne et al. 2022, p. 8), the traditional textbook history of paleontology misses on the opportunity to make the past useful for the future. Even further, the kind of engagement with the past that it transmits might come to stand in the way of cultivating a more complex understanding of the history of the field that could help understand and address a host of contemporary issues, such as specimen repatriations (Howe and Rieppel 2024), parachute science (Thompson 2023), evolving fossil legislations (Liston 2018), and gender gap in the field (Black 2018).

3.4 Beyond the Traditional Textbook History

As conceded at the beginning of this section, relying on a short review of textbook historical accounts to evaluate how the paleontological community has traditionally been engaging with its own past has its limits. Nevertheless, when considered as artifacts recording the kind of historical engagement transmitted to upcoming members in the field, looking at textbook histories help shed some light on the historical reckoning that the paleontological community is currently experiencing. The traditional history of the field that has been transmitted for decades is losing relevance in the face of present challenges. Its benefits for the paleontological community are beginning to be outweighed by its disadvantages. Through this traditional historical engagement, the past ceases to be a helpful resource for the members of the paleontological community to tackle contemporary issues of accessibility to the field and equity within it. On the contrary, this traditional history promotes a vision of the past that might delay, if not even prevent, the widespread recognition across the paleontological community of long-running inequities that affect how the science is being carried out and who stands to benefit from its successes.

Wrestling with the growing concern of insuring that the study of the history of life on Earth actually benefits from and to a greater number of people, the paleontological community is facing the need to abandon one deeply rooted engagement with its own past and to grow a new, more critical one, in the sense defined in the previous section. This transition is inevitably made difficult by (1) an expected attachment to the traditional history of the field and (2) the disruptive and evolving nature of the new histories being written. To achieve this transition in the most productive manner, the paleontological community can count on a renewed historiography of paleontology, which has been working on resituating paleontology within global, economic, social, and cultural histories.

4 A New Historiography of Paleontology

This section introduces a new historiography of paleontology which could assist members of the paleontological community in cultivating an approach to the past better suited to address contemporary challenges, such as increasing accessibility to education and career paths in the field for people with diverse backgrounds and identities (Carter et al. 2022) or defining shared ethical research standards in response to evolving and complex global circumstances as in the case of Myanmar amber (Dunne et al. 2022). This section opens with a brief historiographical overview characterizing the most recent approaches in the history of paleontology and what they propose. Mobilizing the three species of history detailed in Section 2, it proceeds with a discussion of the potential advantages that the paleontological community and its upcoming members could enjoy from engaging with that new critical historiography of paleontology.

4.1 Paleontology's Place in History

Since the second half of the twentieth century, a growing cohort of historians of science has been challenging "the interpretations of an older historiography" of paleontology (Rainger 2008, p. 185), which tended to focus on prominent figures in the field and how their works had played a role in the discovery of the geological past. This historiography included primarily works by paleontologists, such as Zittel's *History of Geology and Palaeontology to the End of the Nineteenth* or Edward's *The Early History of Palaeontology*, both cited in the previous section.

Informed by broader changes in the methodologies and problematics of the history of science, historians of paleontology began to pay closer attention to the cultural and social contexts in which paleontological work had been done. In *The Meaning of Fossils*, Rudwick proposed to show how "each period's interpretation of the meaning of fossils may be an illuminating reflection of that period's view of the natural world" (Rudwick 1985, preface). In *Fossils and Archetypes*, Adrian J. Desmond described his own approach to a study of mid-nineteenth century paleontological debates in the following terms:

> So my strategy, broadly speaking, will be to investigate how far abstruse debates over mammal ancestry or dinosaur stance reflected the cultural context and the social commitment of the protagonists, and as a result to

determine the extent to which ideological influences penetrated palaeontology to shape it at both the conceptual and factual level. (Desmond 1982, p. 17)

While approaches like Desmond's allowed for an exploration of the complex interplay between culture, society, and science in the production of paleontological interpretations, the focus remained primarily, if not exclusively, on the Western context. Also, little attention was being given to the history of collections, fieldwork, preparation techniques, exhibitions, and the broader political context informing them. For instance, in the same study, Desmond intentionally left aside the question of the growth of paleontological collections in mid nineteenth-century London, addressing it only with brief references to the colonial and economic contexts of the period:

> Whether the period saw an increase in the number of fossils shipped to London I do not know; certainly crates seemed to be arriving in Bloomsbury in ever increasing numbers from the colonies (particularly South Africa, New Zealand, and Australia), and my impression is that the "Coal age" vertebrates deposited in Jermyn Street increased considerably, as one might have expected given Britain's predominant mining interests. (14)

Historians of science eventually came to turn their attention more systematically toward things like the "crates" on which Desmond had chosen not to focus. They proceeded to show how labor (Shapin 1989), empires (Petitjean et al. 1992), gender (Schiebinger 1993), spaces (Livingstone 2000), and circulation (Secord 2004) had been playing constitutive roles in the making of scientific knowledge. These "turns" in the general history of science set the foundations for a renewed historiography of paleontology investigating not only cultural and social influences on paleontology, but also paleontology's influence on culture and society. Paleontology certainly has a history, but most importantly for this new historiography, paleontology counts as a historical player.

Investigating the global history of nineteenth-century mammal paleontology, *The Age of Mammals* (2023) by Chris Manias articulates some of the most characteristic perspectives of this new historiography of paleontology. First, it sets out to show that "numerous actors were critical for paleontological work: hunters, agricultural workers, miners, builders, preparators, artists, and others. These people were often subordinate but not powerless" (Manias 2023, p. 382). Second, it investigates the interplay between paleontology and global power dynamics, reminding that a "persistent trend has been that work on fossils followed exploitation and control of territory" (383). Third, it uncovers the legacies of older paleontological ideas in today's popular conceptions of the geologic past, evolution, and the environment. For example:

> Despite constant refrains by paleontologists and evolutionary biologists to understand natural diversity in nonhierarchical manners, common language still refers to mammals as "higher" creatures, a norm in the animal world. References to animals like the rhinoceros and hippopotamus as "prehistoric," and megafauna (both living and extinct) being valued as national or local icons, all persist from these nineteenth-century debates. (384)

Finally, it explains (a) how the field and its community came to organize itself around specific geographical locations and institutional centers, since "large collections developed through taking advantage of extraterritorial and colonial links," and (b) how these "structures of inequality and powers [...] have also persisted" (385).

Chapter 7, consisting in a thematic list of some of the most notable works belonging to this new historiography of paleontology, provides a broader picture of its various perspectives on paleontology's past. But Manias' remarks are representative of a general historiographical approach attentive to the impacts and legacies of paleontology's past, both on the field itself and its present community and on "the current world" (4). This historiography takes seriously the cultural and social influence of paleontological work throughout history and in the present. The history of paleontology becomes a way to better understand a host of pressing issues, from public debates over environmental changes and conservation efforts to the balance of powers that has been shaping the production of paleontological knowledge.

4.2 Some Misunderstandings

A couple of common misunderstandings about such an historiographical approach need to be addressed to ensure a fair evaluation of its potential benefits to the paleontological community.

The broader contextualization of historical actors' deeds, motivations, and ideas inevitably leads to more complex and contrasted portraits of their lives and accomplishments. For this reason, historiographical approaches that investigate an always more complex nexus of circumstances can be seen by some as working on tarnishing the image and reputation of generally well-regarded figures of the past. Discussing how Western paleontologists benefited from imperial endeavors, for example, does not aim at vilifying them. While such historical work is never completely free of moral consideration, its goal is not to separate the "heroes" from the "villains." Instead, it is interested in reminding the part that contingency plays in any lives so as to better understand what made it possible for specific individuals and groups of people to think and act in the way they did. "They were people of their time," a common phrase used to sometimes dismiss historical discourses that delve into what is seen as the unsavory part of the past, is precisely the relationship that historians are curious about: how were people defined by the time they lived in, and how did they make it "theirs"? A critical historiography aims at embracing the complexity and sometimes contradicting aspects of actors and events. It helps understand what made certain actions and practices possible by reconstructing and analyzing how historical actors justified or negotiated them. In this regard, it stands in clear contrast to the process of idealization and simplification that most often characterizes monumental and antiquarian approaches to the past.

A second misunderstanding is that bringing to light the interplay between past scientific work and its broader sociocultural context might amount to a depreciation of science, presenting it as a mere reflection of changing social norms and values. This sort of concern about the effects of historical and sociological analyses of science was at its peak during the infamous "science wars" at the end of the last century. The concern over such discourses chipping away at the credibility of paleontologists' expertise and ability to know about the deep past is particularly legitimate considering that paleontology has been for a long time the target of anti-evolutionist, and creationist movements, especially in the United States (Padian 2009, Laurence 2024). It is clear that the new historiography of paleontology inherits from the deconstructive approaches to science that emerged in the 1970s–1980s (Latour and Wooglar 1979, Shapin and Schaffer 1985). For example, Lukas Rieppel opens his *Assembling the Dinosaur* (2019), a history of dinosaur exhibitions during America's Gilded Age, in thought-provoking fashion:

> The dinosaur is a chimera. Some parts of this complex assemblage are the result of biological evolution. But others are products of human ingenuity, constructed by artists, scientists, and technicians in a laborious process that stretches from the dig site to the naturalist's study and museum's preparation lab. The mounted skeletons that have become such a staple of natural history museums most closely resemble mixed media sculptures, having been cobbled together from a large number of disparate elements that include plaster, steel, and paint, in addition to fossilized bone. (Rieppel 2019, p. 1)

Now, recognizing the complexity of paleontological reconstructions of extinct animals is not the same as reducing them to purely social constructs. Instead, an approach like Rieppel's serves to open a whole field of investigation on the types of work and workers that have been making these reconstructions even possible. Its aim is not depreciative but heuristic: casting a light on the people, practices, and resources that contributed to make the paleontological community and the knowledge it produces about the deep past what they are today.

4.3 Potential Benefits

Mobilizing the three species of history, the new historiography of paleontology can best be described as a form of critical history. Investigating the local and global power relations that have informed the development of paleontology since the nineteenth century as well as uncovering its institutional and cultural legacies, this historiography could serve the present paleontological community, not by giving it new monuments and roots, but by bringing to its attention how the past is playing out in its present. Answering how he hoped his book *The Age of Mammals* could benefit the paleontological community, Manias explained, for example, that:

> The book thinks about how palaeontology was deeply emmeshed with politics, empire and economics throughout its history. This isn't just on the level of its material conduct, but still exists on the level of language – hierarchical thinking, ideas of progress, or notions of particular animal groups having 'dominance.' These are concepts which are rooted in nineteenth-century understandings and ideologies, and are deep in the history of the field. By thinking about how these ideas became rooted, we can start thinking about going beyond them. (Monnin and Manias 2024, p. 55)

As a critical history, the new historiography of paleontology provides the knowledge necessary "to break up and dissolve a part of the past" (Nietzsche 1990, p. 75), which legacy is seen as disserving the present paleontological community and its work.

The main benefit of this critical historiography is to present paleontology as a significant player in the way societies think about, talk about, and act upon the world around them. It reminds readers that the history of paleontology is not something unfolding parallel to the rest of history, or with occasional connections to it, but is actually embedded in it. This way of thinking about the history of paleontology makes members of the paleontological community, past and present, historical actors whose collective actions have been having implications for the world around them. It brings to light the social relevance and responsibility of the paleontological community and its individual members. It shows, through historical case studies, that how and why paleontological work is being conducted affects the world around beyond the intended uses of the knowledge thereby produced. Such an historical perspective could help upcoming members of the paleontological community cultivate an always more acute sense of the social and cultural implications of their work and that of their predecessors.

Another potential benefit of this new critical historiography of paleontology comes from the attention it gives to the diverse ecosystem of people involved in the making of paleontological knowledge. By making room for other historical actors and their contributions, this historiography provides valuable insights into the inner workings of the science and redefines the limits of the paleontological community. As Manias explains, "the study of fossils was not just an intellectual pursuit connected with what we might now call disciplines. Defining, locating, extracting, and analyzing fossils required practical knowledge and expertise" (Manias 2023, p. 8). Casting a light on the historical importance of diverse knowledge, such as indigenous fossil legends and geomythologies (Mayor 2008, 2023, Benoit et al. 2024), labor (Barnett 2020, Elliott 2023, Heumann et al. 2024) and know-how (Brown 2012) in the development of the field could help foster new collaborations while re-appreciating past contributions. Drawing conclusions from her sociological study on the work of fossil preparators, Caitlin Wylie explains, for example, that:

> Paying attention to all the workers in a research community reveals that there's more to the development of scientific knowledge, work, and community than PhDs and publications, and underscores the variety of skills and experiences that science relies on. This more complete view of science is fascinating for its own sake. It also invites more people to participate in research, since anyone can develop these skills, regardless of their success in school or expertise in science. (Wylie 2021, p. 16)

The new historiography of paleontology does not only pay attention to the complexity and diversity of the paleontological community. It also shows that its work has always been intertwined with the development of institutions, such as geological surveys, natural history museums, and university departments, each characterized by their own missions, funding schemes, labor division, practices, and culture. Reminding how institutional contexts have been determining research goals and practices could help upcoming members of the paleontological community evaluate the context of their own work more critically by understanding the historical processes that made it the way it is now.

All of these potential benefits of this new historiography of paleontology amount to cultivating an awareness of paleontology as a social endeavor, dependent on a diverse cast of people, determined by specific forms of social organizing, and affecting the world beyond the limits of its own community. It presents the paleontological worker as an irreducible social player, whose work depends on and affects the world in which it is performed.

Arguing for the value of a historical approach casting a critical gaze on traditional heroes and recovering the actions of a wider ensemble of past actors, Howard Zinn, author of the influential *A People History of the United States*, once stated:

> It's a history, I think, that makes the listener, the reader, the imbiber of history, more of a human being and also more of an active person. If the heroes are the important decision makers, all you have to do as a citizen is to go to the polls every two or four years [...]. But now if you take this other set of heroes, your role as a citizen is not simply to vote but to become an active person in a movement for social justice. (Zinn and Suarez 2019, p. 21–22)

A richer and more complex historical narrative encourages individuals to recognize their roles in the historical process transforming the world in which they live. In the same manner, a more critical engagement with the past of paleontology could encourage members of the paleontological community to think always more intentionally about the social conditions and implications of their work, increasing awareness of issues of accessibility and equity affecting it.

5 Cultivating Critical History

5.1 History: A Luxury?

The previous sections have argued that the paleontological community is currently facing an unprecedented situation as far as its engagement with its own past is concerned. Mobilizing a distinction between three species of history, they have proposed an interpretation of that historical reckoning and introduced historiographical resources to assist in answering it productively. The conclusion reached is that, to best address the variety of issues of accessibility and equity in the field, the paleontological community requires a shift from a mainly monumental/antiquarian engagement with its past toward a more critical one.

But integrating the critical historiography of paleontology to the curriculum of upcoming paleontologists faces multiple obstacles related to funding and resources, personnel and training, as well as forums and time. How much resources, effort, hours, and space should reasonably be allocated to historical considerations in the training of scientists? Wouldn't the addition of historical knowledge, especially of critical nature, risk taking precious time out of the scientific training? The paleontological community is not the only scientific community to have ever faced such questions.

Interestingly, one instructive example to consider when approaching these questions relates to the medical field. In the midst of the Civil Rights movement in the United States, physician and physical anthropologist William Montague Cobb (1904–1990) advocated for the desegregation of hospitals and medical schools. A portion of his efforts centered on raising awareness about the history of Black Americans' contributions to the country's health and their struggles to receive adequate medical care and enjoy equitable opportunities in medical careers (Cobb 1951, 1958). For Cobb, the cultural and legal battles against the segregated establishment in health and medicine needed to include a front on medical education. The next generations of physicians and health professionals would have to be aware of the history of their field, including its segregated past, to secure long-term progress in equitable access to health and medical education. If not, how else could the upcoming members of the medical community be expected to recognize and help redress lingering injustices directly connected to their field?

Cobb made his case for history in medical training in a short piece titled "The Value of History." He argued that teaching medical history "imparts perspective" and "promotes tolerance and modesty" by bringing attention to the contributions of "individuals who were not members of the profession" (Cobb 1961, p. 74). It also serves to remind upcoming members of the medical community that "advances in medicine have not been a thing apart from the rest of human history," making it their responsibility to "understand [their] background and relationships in human society" (74). Despite these benefits, Cobb noted that no medical school in the United States was offering medical history as a required part of the curriculum. While some instructors agreed on its importance and were trying to incorporate some historical perspectives in their teaching, others considered "that the pressures upon curriculum time are such that a course in medical history is a luxury which cannot be afforded" (74). This ambivalence about the value of medical history left Cobb perplexed: "If full and accurate history can contribute so much to the understanding of the disorders of an individual patient, should it not be of equal value in respect to the whole broad field of medicine and its subdivisions?" (74).

In other words, how could such contrasting appreciations for the value of the past coexist within the same field? Even if the context in which Cobb raised this rhetorical question is very different from the one in which the paleontological community is currently evolving, it certainly resonates with the situation in which the paleontological community finds itself regarding its appreciation for the past. If the diligent study of the fossil record allows to better understand the past and present of life on Earth, then how could the diligent study of the historical record of the field be considered a "luxury"? Certainly, as argued in Section 4, the critical study of the historical record of paleontology can offer valuable assistance to the paleontological community in understanding its past and present. Critical history should therefore not be seen as a luxury, expansive and unnecessary, but as an essential contributor to the future advancement of paleontology, both as a science and a social endeavor.

While recognizing the need for critical history in the advancement of paleontology does not solve the problem of how much time, effort, and space should be allocated to it, this recognition does avoid framing the problem of the integration of critical history in paleontology training as if it were a binary choice between (1) more history, less science and (2) less history, more science.

5.2 Growing Historical Awareness

The shift from a primarily monumental/antiquarian engagement with the past to a more critical one does not require cutting into the scientific training. The problem is mostly qualitative in nature, not quantitative. If this shift is to occur, students of paleontology do not necessarily need to engage more with the past of their discipline, but differently.

Integrating the new historiography of paleontology in the training of upcoming paleontologists and in the broader culture of the paleontological community is not a matter of adding required hours dedicated to the history of paleontology. Reflecting on the state of gender inequities in science at the turn of the twenty-first century, historian of science Londa Schiebinger already warned that, "Increasing the numbers of women without increasing an awareness of gender issues will have little impact on science or its institutions" (Schiebinger 1997, p. 210).

Similarly, the focus should not be on increasing the time and space devoted to the history of paleontology, but more so on growing an awareness among current and upcoming members of the paleontological community that present issues of accessibility and equity in the field have an historical component and that the past matters if one hopes to address them effectively for the benefit of the science. The focus should be put on bringing attention to the fact that the history of paleontology can serve as a resource and does not simply constitute a backstory detached from the present. Growing such an historical awareness can be achieved through a variety of initiatives depending on the amount of time and resources available. Even small adjustments can make a difference in cultivating a shared and more critical engagement with the past. The following section lists some suggestions for such initiatives at the levels of instruction, publication, and research.

5.3 Suggestions

To cultivate a more critical approach to the past of the field a variety of initiatives can be contemplated, from minor adjustments to more ambitious projects. These initiatives can be implemented, at least, at three different levels: instruction, publication, and research. What follows are suggestions that might interest readers depending on their activities and available resources. Within each subsection, the suggestions are listed in order of less to more resource-expansive.

5.3.1 Instruction

- Reminding students that authors of historic paleontological works did not work alone and that the publications of the Cuviers and Bucklands have always been supported by a broader cast of workers, from wives and daughters (Kölbl-Ebert 1997) to field workers, guides, and informants.
- Bringing students' attention to the historical and current contexts of significant fossil discoveries and collections to help them realize that access to fossiliferous sites and specimens has been, and continues to be, informed by changing political, legal, and economic circumstances.
- Advertising to students reading suggestions on the history of paleontology to encourage them to see the historiography as a growing resource to situate the field within history and society. Section 7 provides a thematic bibliography that can be used to that effect.
- Inviting speakers, such as historians, archivists, or community leaders, to discuss the history of the field and how it might inform some of its present challenges.
- Building and sharing teaching resources integrating elements of the critical historiography of paleontology. For example, starting in 2020, a group of geoscience educators developed a collection of modules, named GeoContext, "to promote the integration of topics on racism, colonialism, imperialism, environmental damage, and exploitation of natural resources into subjects commonly taught in geoscience departments."[15]

5.3.2 Publication

- Encouraging the review of books and articles pertaining to the history of paleontology in paleontological journals to widen the reception of these works and foster conversations around them.
- Taking advantage of alternative modes of publication besides textbooks and journals, such as newsletters, blogs, and podcasts, to grow forums for interdisciplinary discussions of historical questions. The *Palaeontology Newsletter*, *Priscum*, and *Palaeocast* are examples of such platforms more readily conducive to critical historical discussions.

---

[15] https://serc.carleton.edu/teachearth/geocontext/index.html. Accessed on April 2, 2025.

- Collaborating between historians and paleontologists to write chapters or sections in textbooks and other teaching materials that address the history of paleontology in a more critical manner, as characterized in the previous chapters.
- Collaborating between historians and paleontologists on books or special issues merging the expertise and interests of both disciplines, as in the cases of the *Colligo* special issue "Palaeontological collections in the making" (2020) edited by historians Irina Podgorny and Maria Margaret Lopes, and paleontologist Éric Buffetaut, or the volume *Palaeontology in Public* (2025) edited by Chris Manias and bringing together perspectives from humanities scholars and paleontologists.

5.3.3 Research

- Joining or advertising existing interdisciplinary research groups which share an interest in paleontology, and affiliated sciences, and their history, such as the PopPalaeo (Popularizing Palaeontology: Current & Historical Perspectives) network[16] and INHIGEO (International Commission on the History of Geological Sciences)[17].
- Organizing roundtables and panels at conferences bringing together historians and members of the paleontological community around specific issues, such as fossil legislation, parachute science, or gender equity within the field and how they have affected the science and its social impact.
- Promoting or emulating research projects measuring the effects of historical legacies on the field, such as Pal(a)eoScientometrics[18].
- Collaborating with historians and members of the paleontological community on grant applications for research projects either history-focused, paleontology-focused, or education-focused.

6 Conclusion: Bridging Two Historical Disciplines

Historians of science have long been making the case for the benefits of their work to science (Sarton 1913; Cohen 1955; Heilbron 1987; Gooday et al. 2008; Chang 2017). And it is certainly not the first time that the history of paleontology is advertised to the paleontological community as a valuable resource. Observing the rise of the computer in paleontology around the 1970s, Martin

---

[16] https://www.poppalaeo.com/. Accessed on April 2, 2025.
[17] https://inhigeo.org/. Accessed on April 2, 2025.
[18] https://paleoscientometrics.github.io/. Accessed on April 2, 2025.

Rudwick concluded his history of the interpretation of fossils with a reflection on the importance to remember the rich intellectual history of the field:

> As palaeontology now prepares, therefore, for a great leap forward into a computerised age (for which the nature of its material makes it highly appropriate) there is perhaps a danger that it may lose sight of its historic origins in the 'steam age' of science and before. That it should not become a-historical in outlook is important, not for nostalgic antiquarian reasons, but because the loss of historical perspective would lead to conceptual impoverishment. In every period of its history, palaeontology, like all other branches, has developed through a series of intricate interactions between philosophical presuppositions (often implicit or even unrecognized), theoretical constructions at all levels, and the steadily accumulating fund of observational evidence. An exclusive pre-occupation with the last of these, however excusable it may be in the heat of the present information explosion, will not lead to a more securely and factually based science, but quite probably to a vast superstructure built on unexamined and perhaps weak conceptual foundations. Reflection on the history of palaeontology, with its reminder of the very different worlds of thought in which the science acquired the various strands of its present complex texture, may perhaps help in the critical examination and re-appraisal of its present foundations, and thus ensure that a computerized access to its vast stores of factual information is used to the best heuristic advantage. (Rudwick 1985, p. 266)

The history of paleontology can indeed serve paleontology by clarifying and bringing attention to the "foundations" of its methodologies and theories, thereby assisting in the effort of improvement, adjustment, and correction critical to any scientific endeavor. A significant amount of scholarship continues to provide perspectives, both historical and philosophical, to better understand and assist the production of paleontological knowledge (Turner 2011, Sepkoski 2012, Currie 2018, Tamborini 2020). But this Element's argument is motivated by the desire to put the history of paleontology to work in yet another way: to provide perspectives on the complex social and political history that shaped the paleontological community, along with its institutions, practices, and collections, and that informs present issues of accessibility (e.g. retention of underrepresented and underserved people at every level of academic education and research), ethics (e.g. repatriation of fossil specimens), and equity (e.g. fairness in recognition and attribution of credit) in the field. To be understood, many of these complex issues have to be connected to the past of paleontology, which early development as a field was contemporaneous to and entangled with various colonial enterprises (Manias 2021). This past has left a deep impression on museum fossil collections (Díez Díaz et al. 2025), research practices (Cisneros et al. 2022), and the field's culture (Monarrez et al. 2022).

While the argument of this Element focuses on the importance of the kind of historical engagement needed to address such issues, it does not claim the critical historiography of paleontology to be a panacea. The more intentional and systematic integration of critical perspectives constitutes but one step toward effectively addressing such issues. Nevertheless, this Element stresses the point that this step is an essential one, since the kind of historical engagement a community cultivates with its past inevitably informs its ability to perceive, understand, and act upon the present. Any kind of engagement with the past does not serve or deserve a community in the same ways. For this reason, how the paleontological community decides to proceed in regard to its engagement with its past is a determining factor in how the field will tackle complex problems of accessibility, ethics, and equity for tomorrow's paleontology.

As historical disciplines, the history of paleontology and paleontology share a lot of the same methodological problems on how to know anything about the past and how to make that knowledge relevant for the present (Plotnick et al. 2023). Despite dealing with different pasts, they both experience the frustrations and thrills of working with fragments and traces. They both have to find ways to cautiously account for a multitude of biases in the records of the past. Reflecting on their collaborative effort to revise the history of the Tendaguru expedition and its fossil finds, the authors of *Deconstructing Dinosaurs* point out that:

> While skeletons are assembled from isolated bones, histories are pieced together out of fragmentary records. With every new perspective and every new source having the potential to challenge and alter the interpretation of the data, specimens and narratives alike – including the stories we tell in this book – are temporary constructions, shaped by the institutional, political and social environments that produced them. (Heumann et al. 2024, p. 3)

The temporary nature of historical narratives or reconstructed skeletons from the deep past is neither an argument against their scientific value nor an argument for the equal value of any historical narratives or reconstructed skeletons. Instead, it is a recognition that our best efforts to know anything about the past, human or geological, will always need to be renewed. This sharing in analogous struggles, joys, and ambitions should be seriously considered as a chance to develop meaningful and productive bridges between paleontology and the history of paleontology. So that what and how we learn about the past of life on Earth continue to serve an always greater number of people.

The argument of this Element rests on the belief that the history of paleontology and paleontology have much to learn from each other and that both disciplines can only benefit from cultivating a more intentional and productive relationship with one another. If the present argument insisted on how a new historiography of paleontology could assist the members of the paleontological community, the reciprocal is true as well. As Bernhard Ziegler points out in the preface to his *Introduction to Palaeobiology*, discussed in Section 3, paleontology "is also a historical science that can greatly enrich the Humanities" (Ziegler 1983, p. 9). Members of the paleontological community have access to a wealth of knowledge, material, and experience that historians of paleontology need to orient and pursue their investigation of the field's history. As contemporary witnesses and actors of the most recent transformations of the field, members of the paleontological community can, of course, provide invaluable testimonies. For some of them, as curators of fossil and archival collections, they hold the keys to, and unique perspectives on, materials essential for the work of historians. As students of an even more remote past, they can share wisdom most beneficial to historians. Last but not least, their concerns and questions about the present and future of the field ring to the historian's ear as echoes of a past to recover.

7 Thematic Bibliography

This section is intended to provide reading suggestions for instructors, students, researchers, curators, preparators, and any other members of the paleontological community interested in delving into and putting to use the growing critical historiography of paleontology. The bibliography is by no means exhaustive, leaving aside, for example, important scholarship about the history of geology and other connected sciences (e.g. Yusoff 2018, Bobbette and Donovan 2019, Bashford et al. 2023). Organized under six themes, it proposes instead a selection of key contributions focusing primarily on fossils and paleontology: (1) paleontology, patrons, and institutions; (2) indigenous fossil knowledge; (3) contested fossil heritage; (4) paleontology, nations, and empires; (5) paleontological labor; (6) women in paleontology. Most works could have been cross-listed, but each of them has only been listed under one theme. Besides its focus on works by historians, the bibliography also includes some important contributions from legal scholars, sociologists, anthropologists, and paleontologists. The works in each thematic section are listed in chronological order of publication.

7.1 Paleontology, Patrons, and Institutions

The following works reconnect the development of paleontological practices, methods, knowledge, and collections with the development of museal institutions and state-funded geological surveys, most particularly in the second half of the nineteenth century and at the turn of the twentieth century. Most of them address the exhibition of fossil specimens and the public-facing dimension of paleontology, questioning how institutions, sources of funding, scientific knowledge, politics, and popular culture have been (1) interacting to forge specific visions of the geological past and (2) influencing paleontological research agendas.

- Brinkman, P. (2010). *The Second Jurassic Dinosaur Rush: Museums of Paleontology in America at the Turn of the Twentieth Century*. Chicago, IL: University of Chicago Press.
- Nieuwland, I. (2010). The colossal stranger. Andrew Carnegie and Diplodocus intrude European Culture, 1904–1912. *Endeavour* 34(2): 61–68.
- Rieppel, L. (2012). Bringing dinosaurs back to life: exhibiting prehistory at the American Museum of Natural History. *Isis* 103(3): 460–490.
- Noble, B. (2016). *Articulating Dinosaurs: A Political Anthropology*. Toronto: University of Toronto Press.
- Davidson, J. P. (2017). *Patrons of Paleontology: How Government Support Shaped a Science*. Bloomington, IN: Indiana University Press.
- Rieppel, L. (2019). *Assembling the Dinosaur: Fossil Hunters, Tycoons, and the Making of a Spectacle*. Cambridge, MA: Harvard University Press.
- Nieuwland, I. (2020). *American Dinosaur Abroad: A Cultural History of Carnegie's Plaster Diplodocus*. Pittsburgh, PA: Pittsburgh University Press.

7.2 Indigenous Fossil Knowledge

The readings in this section discuss the existence, historical impact, and erasure of indigenous fossil knowledge in the history of paleontology. They explore the ambivalent relationship that modern paleontology has had with non-scientific, local knowledge about fossils, fossiliferous sites and their origins. They also discuss how this kind of knowledge can be integrated into paleontological research and given full recognition for its cultural value and potential scientific value.

- Mayor, A. (2008). Suppression of Indigenous Fossil Knowledge. From Claverack, New York, 1705 to Agate Springs, Nebraska, 2005. In R. Proctor and L. Schiebinger, eds, *Agnotology: The Making and Unmaking of Ignorance*. Stanford, CA: Stanford University Press, 163-182.
- Xing, L, Mayor, A., Chen, Y. et al. (2011). The folklore of dinosaur trackways in China: impact on paleontology. *Ichnos* 18(4): 213–220.
- Mayor, A. (2023). *Fossil Legends of the First Americans, 2nd edition*. Princeton, NJ: Princeton University Press.
- Benoit, J., Penn-Clarke, C. R., Rust, R. et al. (2024). Indigenous knowledge of palaeontology in Africa. *Geological Society, London, Special Publication* 543: 357–370.

7.3 Contested Fossil Heritage

This section of the bibliography focuses on the dispossession of fossil heritage and disputes over fossil specimens. The following readings raise the question of who is entitled to the care and ownership of fossil specimens. This most important question is addressed from the points of view of legal, national, and international histories. Especially since the beginning of modern paleontology, fossils have been occupying a place at the boundary of different scientific, cultural, economic, and political interests.

- Lazerwitz, D. (1994). Bones of contention: the regulation of paleontological resources on the federal public lands. *Indiana Law Journal* 69(2): 601–636.
- Dussias, A. (1996). Science, sovereignty, and the sacred text: paleontological resources and Native American rights. *Maryland Law Review* 55(1): 84–159.
- Fan, F. (2013). Circulating material objects: the international controversy over antiquities and fossils in twentieth-century China. In B. Lightman et al., eds. *The Circulation of Knowledge Between Britain, India and China*. Leiden: Brill, 209–236.
- Bradley, L. (2014). *Dinosaurs and Indians: Paleontology Resource Dispossession From Sioux*. Parker, CO: Outskirts Press.
- Jones, E. D. (2020). Assumptions of authority: the story of Sue the *T-rex* and controversy over access to fossils. *History and Philosophy of the Life Sciences* 42(1): 1–27.
- Yen, H. (2024). Fossils and sovereignty: science diplomacy and the politics of deep time in the Sino-American fossil dispute of the 1920s. *Isis* 115(1): 1–22.

7.4 Paleontology, Nations, and Empires

The ambition shared by the works in this section is to situate the development of paleontological knowledge and collections within broader histories of nations, empires, and global circulations. On one hand, they discuss how the collection and interpretation of fossils have been informed by nation-building processes, imperial endeavors, trading routes, and information networks. On the other hand, they explain how fossils have been playing a role, not only in revealing the deep history of life on Earth, but also in formulating visions of modernity.

- Semonin, P. (2000). *American Monster: How the Nation's First Prehistoric Creature Became a Symbol of National Identity*. New York, NY: New York University Press.

- Podgorny, I. (2013). Fossil dealers, the practices of comparative anatomy and British diplomacy in Latin America, 1820–1840. *The British Journal for the History of Science* 46(4): 647–674.
- Tamborini, M. (2016). "If the American can do it, so can we": how dinosaur bones shaped German paleontology. *History of Science* 54(3): 225–256.
- Pimentel, J. (2017). *The Rhinoceros and the Megatherium: An Essay in Natural History.* Cambridge, MA: Harvard University Press.
- Heumann, I., Vennen, M. and Stoecker, H. (eds.) (2024). *Deconstructing Dinosaurs: The History of the German Tendaguru Expedition and its Finds, 1906–2023.* Leiden: Brill.
- Zizzamia, D. (2019). Restoring the paleo-West: fossils, coal, and climate in late-nineteenth century America. *Environmental History* 24(1): 130–156.
- Chakrabarti, P. (2020). *Inscriptions of Nature: Geology and the Naturalization of Antiquity.* Baltimore, MD: Johns Hopkins University Press.
- Manias, C. (2021). Colonialism and palaeontology: connected histories. *The Palaeontology Newsletter* 106: 59–62.
- Manias, C. (2023). *The Age of Mammals: Nature, Development and Paleontology in the Long Nineteenth Century.* Pittsburgh, PA: Pittsburgh University Press.
- Rieppel, L. and Chang, Y. (2023). Locating the Central Asiatic expedition: epistemic imperialism in vertebrate paleontology. *Isis* 114(4): 725–746.
- Winterer, C. (2024). *How the New World Became Old: The Deep Time Revolution in America.* Princeton, NJ: Princeton University Press.
- Qureshi, S. (2025). *Vanished: An Unnatural History of Extinction.* London: Allen Lane.

7.5 Paleontological Labor

These studies explore the multiplicity of actors that have been contributing to paleontology. While other works listed in this bibliography also touch upon this question, the ones listed in this section focus on the ways in which different categories of actors have been collaborating, negotiating, and crediting (or not) each other. These questions are approached in the contexts of fieldwork, preparation, artistic reconstruction, illustration, and publication.

- Nair, S. P. (2005). "Eyes and no eyes": Siwalik fossil collecting and the crafting of Indian palaeontology (1830–1847). *Science in Context* 18(3): 359–392.
- Vetter J. (2008). Cowboys, scientists, and fossils: The field site and local collaboration in the American West. *Isis* 99(2): 273–303.

- Brown, M. (2012). The development of "modern" palaeontological laboratory methods: a century of progress. *Earth and Environmental Science Transactions of the Royal Society of Edinburgh* 103(3–4): 205–216.
- Rieppel, L. (2015). Prospecting for dinosaurs on the mining frontier: the value of information in America's Gilded Age. *Social Studies of Science* 45(2): 161–186.
- Barnett, L. (2020). Showing and hiding: the flickering visibility of earth workers in the archives of earth science. *History of Science* 58(3): 245–274.
- Witton, M. and Michel, E. (2022). *Art and Science of the Crystal Palace Dinosaurs*. Wiltshire, UK: The Crowood Press.
- Wylie, C. D. (2021). *Preparing Dinosaurs: The Work Behind the Scenes*. Cambridge, MA: MIT Press.

7.6 Women in Paleontology

The publications in this section share two objectives: (1) bringing to light the contributions of women to paleontology throughout history and (2) understanding how these contributions had been erased or forgotten. Members of the paleontological and geological communities have been very active in recovering the history of women who collected, described, interpreted, and illustrated fossils. Among the publications listed, the volume by Annalisa Berta and Susan Turner constitutes, to date, the most extensive survey of women who worked on vertebrate paleontology within the last 200 years. Much work remains to be done to recover the stories of women's accomplishments in all branches of paleontology.

- Aldrich, M. (1982). Women in paleontology in the United States, 1840–1960. *Earth Sciences History* 1(1): 14–22.
- Torrens, H. (1995). Mary Anning (1799–1847) of Lyme; 'the greatest fossilist the world ever knew.' *The British Journal for the History of Science* 28(3): 257–284.
- Turner, S., Burek, C. V. and Moody, R. T. J. (2010). Forgotten women in an extinct saurian (man's) world. *Geological Society, London, Special Publications* 343: 111–153.
- Kölbl-Ebert, M. (2012). Sketching rocks and landscape: drawing as a female accomplishment in the service of geology. *Earth Sciences History* 31(2): 270–286.
- Pickford, S. (2015). "I have no pleasure in collecting for myself alone": social authorship, networks of knowledge and Etheldred Benett's *Catalogue of the Organic Remains of the County of Wiltshire* (1831). *Journal of Literature and Science* 8(1): 69–85.

- Gries, R. R. (2018a). How female geologists were written out of history: the micropaleontology breakthrough. In B. Johnson, ed., *Women and Geology: Who Are We, Where Have We Come From, and Where Are We Going?*. Boulder, CO: The Geological Society of America 214: 11–21.
- Gries, R. R. (2018b). *Anomalies: Pioneering Women in Petroleum Geology, 1917-2017*. Revised Edition. Longmont: Steuben Press.
- Berta, A. and Turner, S. (2020). *Rebels, Scholars, Explorers: Women in Vertebrate Paleontology*. Baltimore, MD: Johns Hopkins University Press.

References

Adams, F. D. (1938). *The Birth and Development of the Geological Sciences*. Baltimore, MD: Williams & Wilkins.

Aldrich, M. (1982). Women in paleontology in the United States, 1840–1960. *Earth Sciences History* 1(1): 14–22.

Andrews, H. N. (1980). *The Fossil Hunters: In Search of Ancient Plants*. Ithaca, NY: Cornell University Press.

Armstrong, E. S. and Sharp, C.-M. (2024). Editorial: Mobilizing museum minerals. *Museum & Society* 22(2–3): 1–14.

Bakker, R. T. (1986). *Dinosaur Heresies: New Theories Unlocking the Mystery of the Dinosaurs and their Extinction*. New York, NY: William Morrow and Co.

Barnett, L. (2020). Showing and hiding: The flickering visibility of earth workers in the archives of earth science. *History of Science* 58(3): 245–274.

Bashford, A., Kern, E. and Bobbette A. (eds.) (2023). *New Earth Histories: Geo-Cosmologies and the Making of the Modern World*. Chicago, IL: Chicago University Press.

Benoit, J. (2018). What would it mean to decolonize palaeontology? Here are some ideas. *The Conversation*: https://theconversation.com/what-would-it-mean-to-decolonise-palaeontology-here-are-some-ideas-102133. Accessed on August 12, 2025.

Benoit, J., Penn-Clarke, C. R., Rust, R. et al. (2024). Indigenous knowledge of palaeontology in Africa. *Geological Society, London, Special Publication* 543: 357–370.

Benton, M. J. and Harper, D. A. T. (2009). *Introduction to Paleobiology and the Fossil Record*. Hoboken, NJ: Wiley-Blackwell.

Bernard, R. E. and Cooperdock, E. H. G. (2018). No progress on diversity in 40 years. *Nature Geoscience* 11: 292–295.

Berta, A. and Turner, S. (2020). *Rebels, Scholars, Explorers: Women in Vertebrate Paleontology*. Baltimore, MD: Johns Hopkins University Press.

Black, R. (2018). The many ways women get left out of paleontology. *Smithsonian Magazine*.

Bobbette, A. and Donovan, A. (eds.) (2019). *Political Geology: Active Stratigraphies and the Making of Life*. Cham: Palgrave Macmillan.

Boule, M. and Piveteau, J. (1935). *Les Fossiles: Éléments de Paléontologie*. Paris: Masson & Cie.

Bradley, L. (2014). *Dinosaurs and Indians: Paleontology Resource Dispossession From Sioux*. Parker, CO: Outskirts Press.

Brinkman, P. (2010). *The Second Jurassic Dinosaur Rush: Museums of Paleontology in America at the Turn of the Twentieth Century*. Chicago, IL: University of Chicago Press.

Brobjer, T. H. (2004). Nietzsche's view of the value of historical studies and methods. *Journal of the History of Ideas* 65(2): 301–322.

Brown, M. (2012). The development of "modern" palaeontological laboratory methods: A century of progress. *Earth and Environmental Science Transactions of the Royal Society of Edinburgh* 103(3–4): 205–216.

Buckland, W. (1824). Notice on the Megalosaurus or great Fossil Lizard of Stonesfield. *Transactions of the Geological Society* 1(2): 390–396.

Buffetaut, E. (1987). *A Short History of Vertebrate Palaeontology*. London: Croom Helm.

Carter, A. M., Johnson, E. H. and Schroeter, E. R. (2022). Long-term retention of diverse paleontologists requires increasing accessibility. *Frontiers in Ecology and Evolution* 10: 876806.

Chakrabarti, P. (2020). *Inscriptions of Nature: Geology and the Naturalization of Antiquity*. Baltimore, MD: Johns Hopkins University Press.

Chang, H. (2017). Who cares about the history of science? *Notes and Records* 71(1): 91–107.

Chiarella, D. and Vurro, G. (2020). Fieldwork and disability: An overview for an inclusive experience. *Geological Magazine* 157(11): 1933–1938.

Cisneros, J. C., Raja, N. B., Ghilardi, A. M. et al. (2022). Digging deeper into colonial palaeontological practices in modern day Mexico and Brazil. *Royal Society Open Science* 9(3): 210898.

Clark, D. L. (1968). *Fossils, Paleontology and Evolution*. Dubuque, IA: Brown.

Cobb, W. M. (1951). The Negro nurse and the nation's health. *The Journal of Negro Education* 20(1): 126–130.

  (1958). Not to the swift: Progress and prospects of the Negro in science and the professions. *The Journal of Negro Education* 27(2): 120–126.

  (1961). The value of medical history. *Negro History Bulletin* 24(4): 74, 90.

Cohen, I. B. (1955). Present status and needs of the history of science. *Proceedings of the American Philosophical Society* 99(5): 343–347.

Colbert, E. H. (1955). *Evolution of the Vertebrates: A History of the Backboned Animals through Time*. New York, NY: Wiley.

  (1968). *Men and Dinosaurs: The Search in Field and Laboratory*. New York, NY: Dutton & Co.

References

Croneis, C. (1942). Reviewed Work: *A Catalogue of the Foraminifera* by Brooks F. Ellis and Angelina R. Messina. *The Journal of Geology* 50(5): 560–562.

Currie, A. (2018). *Rock, Bone, and Ruin: An Optimist's Guide to the Historical Sciences*. Cambridge, MA: The MIT Press.

Darwin, C. (1840). *Journal of Research into the Geology and Natural History of the Various Countries visited by H.M.S. Beagle, under the Command of Captain Fitzroy, R.N. from 1832 to 1836*. London: Henry Colburn.

Das, S. and Lowe, M. (2018). Read in Black and White: Decolonial approaches to interpreting natural history collections. *Journal of Natural Science Collections* 6: 4–14.

Daston, L. (2012). The sciences of the archive. *Osiris* 21: 156–187.

Davidson, J. P. (2017). *Patrons of Paleontology: How Government Support Shaped a Science*. Bloomington, IN: Indiana University Press.

Díez Díaz, V., Akhlaq, S., Kaiser, K. et al. (2025). Digitization as a research methodology in colonial natural history collections. *Nature Reviews Biodiversity* 1: 145–146.

Dunne, E. M., Raja, N. B., Stewens, P. P. et al. (2022). Ethics, law, and politics in palaeontological research: The case of Myanmar amber. *Communications Biology* 5: 1023.

Dunne, E. M., Chattopadhyay, D., Dean, C. D. et al. (2025). Data equity in paleobiology: Progress, challenges, and future outlook. *Paleobiology* 51: 237–249.

Dussias, A. (1996). Science, sovereignty, and the sacred text: Paleontological resources and Native American rights. *Maryland Law Review* 55(1): 84–159.

Edwards, W. N. (1967). *The Early History of Palaeontology*. London: The British Museum.

Elliott, C. (2023) The first fossil finders in North America Were enslaved and Indigenous people. *Smithsonian Magazine*, February 22: www.smithsonianmag.com/history/the-first-fossil-finders-in-north-america-were-enslaved-and-indigenous-people-180981615/?no-cache. Accessed on August 5, 2025.

Fan, F. (2013). Circulating material objects: The international controversy over antiquities and fossils in twentieth-century China. In B. Lightman et al., eds. *The Circulation of Knowledge Between Britain, India and China*. Leiden: Brill, 209–236.

Gooday, G., Lynch, J. M., Wilson, K. G. et al. (2008). Does science education need the history of science? *Isis* 99(2): 322–330.

Gould, S. J. (1987). *Time's Arrow, Time's Cycle*. Cambridge, MA: Harvard University Press.

Gries, R. R. (2018a). How female geologists were written out of history: The micropaleontology breakthrough. In B. Johnson, (ed.), *Women and Geology: Who Are We, Where Have We Come From, and Where Are We Going?*. Boulder, CO: The Geological Society of America 214: 11–21.

(2018b). *Anomalies: Pioneering Women in Petroleum Geology, 1917–2017*. Revised Edition. Longmont, CO: Steuben Press.

Halliday, T. (2022). *Otherlands: A Journey through Earth's Extinct Worlds*. New York, NY: Random House.

Harris, M. (2019). Science wars: The next generation. *Science for the People* 22(1): https://magazine.scienceforthepeople.org/vol22-1/science-wars-the-next-generation/.

Heilbron, J. L. (1987). Applied history of science. *Isis* 78(4): 552–563.

Heumann, I., Vennen, M. and Stoecker, H. (eds.) (2024). *Deconstructing Dinosaurs: The History of the German Tendaguru Expedition and its Finds, 1906–2023*. Leiden: Brill.

Hide, L. (2024). Darwin's chalcopyrite: Engaging museum audiences with global extractive stories. *Museum & Society* 22(3): 162–175.

Hobsbawm, E. J. (1972). The social function of the past: Some questions. *Past & Present* 55: 3–17.

Howe, C. and Rieppel, L. (2024). Why museums should repatriate fossils. *Nature* 630: 559–562.

Hurst, S., Moore, M. W., Simpson, A. et al. (2024). More than museums: Care for natural and cultural heritage in Australia. *Geoconservation Research* 7(2): https://doi.org/10.57647/gcr-2024-si-sy25.

Jones, E. D. (2020). Assumptions of authority: The story of Sue the *T-rex* and controversy over access to fossils. *History and Philosophy of the Life Sciences* 42(1): 1–27.

Kempf, H. L., Olson, H. C., Monarrez, P. M. et al. (2023). History of Native American land and natural resource policy in the United States: Impacts on the field of paleontology. *Paleobiology* 49(2): 191–203.

Kingsbury, C. G., Sibert, E. C., Killingback, Z. et al. (2020). "Nothing about us without us:" The perspectives of autistic geoscientists on inclusive instructional practices in geoscience education. *Journal of Geoscience Education* 68(4): 302–310.

Klein, H. (2020). Protests over George Floyd's death spread around the world. *Slate*: https://slate.com/news-and-politics/2020/06/george-floyd-worldwide-protests.html. Accessed on August 18, 2025.

Kölbl-Ebert, M. (1997). Mary Buckland (née Morland) 1797–1857. *Earth Sciences History* 16(1): 33–38.

(2012). Sketching rocks and landscape: Drawing as a female accomplishment in the service of geology. *Earth Sciences History* 31(2): 270–286.

Latour, B. and Wooglar, S. (1979). *Laboratory Life: The Social Construction of Scientific Facts*. London: Sage.

Laurence, A. (2024) Of dinosaurs and intergenerational culture wars: Dinomania, nostalgia, and the 'missionary lizards' of young earth creationism. *Interdisciplinary Science Reviews* 49(3–4): 389–409.

Lazerwitz, D. (1994). Bones of contention: The regulation of paleontological resources on the federal public lands. *Indiana Law Journal* 69(2): 601–636.

Lister, A. (2018). *Darwin's Fossils: The Collection that Shaped the Theory of Evolution*. Smithsonian Books.

Liston, J. (2018). Guest editorial. *Geological Curator* 10(10): 549–550.

Livingston, D. N. (2000). Making space for science. *Erdkunde* 54(4): 285–296.

MacLeod, N. (2006). Review of Mayor, Adrienne, *Fossil Legends of First Americans*. *Palaeontologia Electronica* 9(2): R4.

Manias, C. (2021). Colonialism and palaeontology: connected histories. *The Palaeontology Newsletter* 106: 59–62.

(2023). *The Age of Mammals: Nature, Development, & Paleontology in the Long Nineteenth Century*. Pittsburgh, PA: University of Pittsburgh Press.

(ed.) (2025). *Palaeontology in Public: Popular Science, Lost Creatures and Deep Time*. London: UCL Press.

Mantell, G. (1847). *Geological Excursions Round the Isle of Wight*. London: Henry G. Bohn.

Marsh, L. J. and Currano, E. (eds.) (2020). *The Bearded Lady Project*. New York, NY: Columbia University Press.

Mayor, A. (2008). Suppression of Indigenous fossil knowledge. From Claverack, New York, 1705 to Agate Springs, Nebraska, 2005. In R. Proctor and L. Schiebinger, eds., *Agnotology: The Making and Unmaking of Ignorance*. Stanford, CA: Stanford University Press, 163–182.

(2023). *Fossil Legends of the First Americans*. 2nd edition. Princeton, NJ: Princeton University Press.

Monarrez, P. M., Zimmt, J. B., Clement, A. M. et al. (2022). Our past creates our present: A brief overview of racism and colonialism in Western paleontology. *Paleobiology* 48(2): 173–185.

Monnin, V. and Manias, C., (2024). Interview with Chris Manias, author of *The Age of Mammals*. *The Palaeontology Newsletter* 115: 53–55.

Nair, S. P. (2005). "Eyes and no eyes": Siwalik fossil collecting and the crafting of Indian palaeontology (1830–1847). *Science in Context* 18(3): 359–392.

Olcott, A. N. and Downen, M. R. (2020). The challenges of fieldwork for LGBTQ+ geoscientists. *Eos* 101: https://doi.org/10.1029/2020EO148200.

Nietzsche, F. (1997). *Untimely Meditations.* Cambridge, UK: Cambridge University Press.

Nieuwland, I. (2010). The colossal stranger. Andrew Carnegie and Diplodocus intrude European Culture, 1904–1912. *Endeavour* 34(2): 61–68.

Nieuwland, I. (2020). *American Dinosaur Abroad: A Cultural History of Carnegie's Plaster Diplodocus.* Pittsburgh, PA: Pittsburgh University Press.

Noble, B. (2016). *Articulating Dinosaurs: A Political Anthropology.* Toronto: University of Toronto Press.

Osborn, H. F. (1931). *Cope: Master Naturalist.* Princeton, NJ: Princeton University Press.

Padian, K. (2009). The evolution of creationists in the United States: where are they now, and where are they going? *Comptes Rendus Biologies* 332(2–3): 100–109.

Panciroli, E. (2017). Beards and Gore-Tex: does palaeontology have an image problem? *The Guardian*: www.theguardian.com/science/2017/aug/16/beards-and-gore-tex-does-palaeontology-have-an-image-problem. Accessed on August 15, 2025.

Petitjean, P., Jami, C. and Moulin, A. M. (eds.) (1992). *Science and Empires: Historical Studies about Scientific Development and European Expansion.* Dordrecht: Kluwer.

Pickford, S. (2015). "I have no pleasure in collecting for myself alone": social authorship, networks of knowledge and Etheldred Benett's *Catalogue of the Organic Remains of the County of Wiltshire* (1831). *Journal of Literature and Science* 8(1): 69-85.

Pimentel, J. (2017). *The Rhinoceros and the Megatherium: An Essay in Natural History.* Cambridge, MA: Harvard University Press.

Plotnick, R. E., Anderson, B. M., Carlson, S. J. et al. (2023). Paleontology is far more than new fossil discoveries. *Scientific American*: www.scientificamerican.com/article/paleontology-is-far-more-than-new-fossil-discoveries1/. Accessed on April 19, 2025.

Podgorny, I. (2013). Fossil dealers, the practices of comparative anatomy and British diplomacy in Latin America, 1820–1840. *The British Journal for the History of Science* 46(4): 647–674.

Podgorny, I., Buffetaut, E. and Lopes, M. M. (2020). Paleontological collections in the making – an introduction to the special issue. *Colligo* 3(3): https://perma.cc/L772-PP5T.

Prothero, D. R. (1998). *Bringing Fossils to Life: An Introduction to Paleobiology*. Boston, MA: McGraw-Hill.

Qureshi, S. (2025). *Vanished: An Unnatural History of Extinction*. London: Allen Lane.

Raja, N. B., Dunne, E. M., Matiwane, A. et al. (2022). Colonial history and global economics distort our understanding of deep-time biodiversity. *Nature Ecology & Evolution* 6: 145–154.

Rieppel, L. (2012). Bringing dinosaurs back to life: exhibiting prehistory at the American Museum of Natural History. *Isis* 103(3): 460–490.

(2015). Prospecting for dinosaurs on the mining frontier: the value of information in America's Gilded Age. *Social Studies of Science* 45(2): 161–186.

(2019). *Assembling the Dinosaur: Fossil Hunters, Tycoons, and the Making of a Spectacle*. Cambridge, MA: Harvard University Press.

Rieppel, L. and Chang, Y. (2023). Locating the Central Asiatic expedition: epistemic imperialism in vertebrate paleontology. *Isis* 114(4): 725–746.

Rudwick, M. J. S. (1985). *The Meaning of Fossils: Episodes in the History of Palaeontology*. 2nd edition. Chicago, IL: The University of Chicago Press.

Sarton, G. (1913). L'histoire de la science. *Isis* 1(1): 3–46.

(1949). La transmission au monde modern de la science ancienne et médiévale. *Revue d'histoire des sciences et de leurs applications* 2(2): 101–138.

Schiebinger, L. (1993). *Nature's Body: Gender in the Making of Modern Science*. Boston: Beacon Press.

(1997). Creating Sustainable Science. *Osiris* 12: 201–216.

Schuchert, C. and Dunbar, C. O. (1941). *A Textbook of Geology. Part II – Historical Geology*. 4th edition. New York, NY: Wiley.

Secord, J. A. (2004). Knowledge in transit. *Isis* 95(4): 654–672.

Semonin, P. (2000). *American Monster: How the Nation's First Prehistoric Creature Became a Symbol of National Identity*. New York, NY: New York University Press.

Sepkoski, D. (2012). *Rereading the Fossil Record: The Growth of Paleobiology as an Evolutionary Discipline*. Chicago, IL: Chicago University Press.

Shapin, S. and Schaffer, S. (1985). *Leviathan and the Air-Pump: Hobbes, Boyle, and the Experimental Life*. Princeton, NJ: Princeton University Press.

Shapin, S. (1989). The invisible technician. *American Scientist* 77(6): 554–563.

Simpson, G. G. (1942). The beginnings of vertebrate paleontology in North America. *Proceedings of the American Philosophical Society* 86(1): 130–188.

(1943). The discovery of fossil vertebrates in North America. *Journal of Paleontology* 17(1): 26–38.

Stearn, C. and Carroll, R. (1989). *Paleontology: The Record of Life*. New York, NY: Wiley.

Stiglitz, J. E. (2022). Inequality got much worse. *Scientific American* 326(3): 55.

Stimpson, C. M., Jukar, A. M., Bonea, A. et al. (2024). A 'large and valuable' Siwalik fossil collection in the archives of the Oxford University Museum of Natural History. *Historical Biology* 36(7): 1167–1179.

Stokes, W. L. (1960). *Essentials of Earth History: An Introduction to Historical Geology*. Englewood Cliffs: Prentice-Hall.

(1982). *Essentials of Earth History: An Introduction to Historical Geology*. 4th edition. Englewood Cliffs, NJ: Prentice-Hall.

Stroup, A. (2017). Picturing the Past Through Scientific Illustration. *Field Museum Blog*: https://www.fieldmuseum.org/blog/picturing-past-through-scientific-illustration. Accessed on April 7, 2025.

Tamborini, M. (2016). "If the American can do it, so can we": how dinosaur bones shaped German paleontology. *History of Science* 54(3): 225–256.

(2020). Technoscientific approaches to deep time. *Studies in History and Philosophy of Science Part A* 79: 57–67.

Taquet, P. (1998). *Dinosaur Impressions: Postcards from a Paleontologist*. Trans. by Kevin Padian. Cambridge, UK: Cambridge University Press.

(2006). *Georges Cuvier: Naissance d'un genie*. Paris: Odile Jacob.

(2019). *Georges Cuvier: Anatomie d'un naturaliste*. Paris: Odile Jacob.

Thompson, H. (2023). Paleontology has a 'parachute science' problem. Here is how it plays out in 3 nations. *Science News* 205(5).

Torrens, H. (1995). Mary Anning (1799–1847) of Lyme; 'the greatest fossilist the world ever knew.' *The British Journal for the History of Science* 28(3): 257–284.

Turner, D. (2011). *Paleontology: A Philosophical Introduction*. Cambridge, UK: Cambridge University Press.

(2019). *Paleoaesthetics and the Practice of Paleontology*. Cambridge, UK: Cambridge University Press.

Turner, S., Burek, C. V. and Moody, R. T. J. (2010). Forgotten women in an extinct saurian (man's) world. *Geological Society, London, Special Publications* 343: 111–153.

Valenzuela-Toro, A. M., Viglino, M. and Loch, C. (2025). Historical and ongoing inequities shape research visibility in Latin American aquatic mammal paleontology. *Communications Biology* 8: 472.

Vetter, J. (2008). Cowboys, scientists, and fossils: The field site and local collaboration in the American West. *Isis* 99(2): 273–303.

Vitaliano, D. B. (1973). *Legends of the Earth: Their Geologic Origins*. Bloomington, IN: Indiana University Press.

Weller, J. M. (1960). Development of Paleontology. *Journal of Paleontology* 34(5): 1001–1019.

Winterer, C. (2024). *How the New World Became Old: The Deep Time Revolution in America*. Princeton, NJ: Princeton University Press.

Witton, M. and Michel, E. (2022). *Art and Science of the Crystal Palace Dinosaurs*. Wiltshire, UK: The Crowood Press.

Wylie, C. D. (2021). *Preparing Dinosaurs: The Work Behind the Scenes*. Cambridge, MA: MIT Press.

——— (2021). What fossil preparators can teach us about more inclusive science. *Issues in Science and Technology* 38(1): 14-16.

Xing, L, Mayor, A., Chen, Y. et al. (2011). The folklore of dinosaur trackways in China: impact on paleontology. *Ichnos* 18(4): 213–220.

Yen, H. (2024). Fossils and sovereignty: science diplomacy and the politics of deep time in the Sino-American fossil dispute of the 1920s. *Isis* 115(1): 1–22.

Yusoff, K. (2018). *A Billion Black Anthropocenes or None*. Minneapolis: University of Minnesota Press.

Ziegler, B. (1983). *Introduction to Palaeobiology: General Palaeontology*. Chichester: Horwood.

Zinn, H. and Suarez, R. (2019). *Truth Has a Power of Its Own*. New York, NY: The New Press.

Zittel (von), K. A. (1901). *History of Geology and Palaeontology to the End of the Nineteenth Century*. London: Walter Scott.

Zizzamia, D. (2019). Restoring the paleo-West: fossils, coal, and climate in late-nineteenth century America. *Environmental History* 24(1): 130-156.

Acknowledgments

I am indebted to Pedro Monarrez for having invited me to present my research on the historiography of paleontology at the 2023 Paleontological Society Short Course on diversity, equity, and inclusion. My gratitude also goes to Brenda Hunda for her guidance throughout the drafting and editing process of this Element. Thank you to Kayleigh Bohémier for sharing very useful insights and information. The thoughtful feedback offered by the two anonymous reviewers were invaluable and gave me a lot to consider for this publication and future projects. The thoughts shared in this Element owe a lot to the inspiring exchanges made possible by the Popularizing Palaeontology network (PopPalaeo) curated by Chris Manias. My sincere thanks to all its members with whom I have been having the pleasure to continue learning about the history of paleontology. Finally, I want to thank my wife and daughter for always being there for me.

Cambridge Elements

Elements of Paleontology

Editor-in-Chief

Brenda R. Hunda
Cincinnati Museum Center

About the Series

The Elements of Paleontology series is a publishing collaboration between the Paleontological Society and Cambridge University Press. The series covers the full spectrum of topics in paleontology and paleobiology, and related topics in the Earth and life sciences of interest to students and researchers of paleontology.

The Paleontological Society is an international nonprofit organization devoted exclusively to the science of paleontology: invertebrate and vertebrate paleontology, micropaleontology, and paleobotany. The Society's mission is to advance the study of the fossil record through scientific research, education, and advocacy. Its vision is to be a leading global advocate for understanding life's history and evolution. The Society has several membership categories, including regular, amateur/avocational, student, and retired. Members, representing some 40 countries, include professional paleontologists, academicians, science editors, Earth science teachers, museum specialists, undergraduate and graduate students, postdoctoral scholars, and amateur/avocational paleontologists.

Cambridge Elements

Elements of Paleontology

Elements in the Series

Phylogenetic Comparative Methods: A User's Guide for Paleontologists
Laura C. Soul and David F. Wright

Expanded Sampling Across Ontogeny in Deltasuchus motherali (Neosuchia, Crocodyliformes): Revealing Ecomorphological Niche Partitioning and Appalachian Endemism in Cenomanian Crocodyliforms
Stephanie K. Drumheller, Thomas L. Adams, Hannah Maddox and Christopher R. Noto

Testing Character Evolution Models in Phylogenetic Paleobiology: A Case Study with Cambrian Echinoderms
April Wright, Peter J. Wagner and David F. Wright

The Taphonomy of Echinoids: Skeletal Morphologies, Environmental Factors and Preservation Pathways
James H. Nebelsick and Andrea Mancosu

Follow The Fossils: Developing Metrics For Instagram As A Natural Science Communication Tool
Samantha B. Ocon, Lisa Lundgren, Richard T. Bex II, Jennifer E. Bauer, Mary Janes Hughes, and Sadie M. Mills

Niche Evolution and Phylogenetic Community Paleoecology of Late Ordovician Crinoids
Selina R. Cole and David F. Wright

Molecular Paleobiology of the Echinoderm Skeleton
Jeffrey R. Thompson

A Review of Blastozoan Echinoderm Respiratory Structures
Sarah L. Sheffield, Maggie R. Limbeck, Jennifer E. Bauer, Stephen A. Hill, and Martina Nohejlová

A Review and Evaluation of Homology Hypotheses in Echinoderm Paleobiology
Colin D. Sumrall, Sarah L. Sheffield, Jennifer E. Bauer, Jeffrey R. Thompson, and Johnny A. Waters

The Ecology of Biotic Interactions in Echinoids: Modern Insights into Ancient Interactions
Elizabeth Petsios, Lyndsey Farrar, Shamindri Tennakoon, Fatemah Jamal, Roger W. Portell, Michał Kowalewski, and Carrie L. Tyler

What Does Graptolite Origination and Extinction Reveal about the Cause of the Late Ordovician Mass Extinction?
Charles E. Mitchell, H. David Sheets, Michael J. Melchin and Chris Holmden

Critical History for Tomorrow's Paleontology
Victor Monnin

A full series listing is available at: www.cambridge.org/EPLY

For EU product safety concerns, contact us at Calle de José Abascal, 56–1°,
28003 Madrid, Spain or eugpsr@cambridge.org.

www.ingramcontent.com/pod-product-compliance
Ingram Content Group UK Ltd.
Pitfield, Milton Keynes, MK11 3LW, UK
UKHW022258240426
470365UK00007B/123